Messias Alessandro Cardoso

From the dream of freedom to migration for capital

Messias Alessandro Cardoso

From the dream of freedom to migration for capital

Forced migration in the sugarcane fields of Gastão Vidigal - SP

ScienciaScripts

Imprint

Cover image: www.ingimage.com

This book is a translation from the original published under ISBN 978-3-330-75674-8.

Publisher:
Sciencia Scripts
is a trademark of
Dodo Books Indian Ocean Ltd. and OmniScriptum S.R.L publishing group

120 High Road, East Finchley, London, N2 9ED, United Kingdom
Str. Armeneasca 28/1, office 1, Chisinau MD-2012, Republic of Moldova, Europe
Printed at: see last page
ISBN: 978-620-8-28089-5

SUMMARY

Procession of the Retirantes

Pedro Munhoz

Terra Brasilis, continent,
Motherland of my people
Today I want to ask you
If your arms are so big, why do you deny space to those who want a home?
I can't understand
that in this immense nation
it's still kill or be killed
For a piece of land
Farmers on the roads

Seeing the land abandoned
with no one to plant
Between fences and fences go millions condemned
to die or beg

I can't understand
To find the clear reason for
those who only live to have
And still call themselves good Christians

In the eldorado of Parà
Carajás Indian name,
the massacre happened
In this land of massacres
these murderous bullets
everyone knows where they come from
Justice and equality need to
be more than just
occasional words
A new era is needed in which it's not just a promise to
share even bread
This is the time to make the division

I can't understand
that instead of inheriting a share,
your people deserve to have
only seven palms of ground

New wave of immigrants
Procession of retreatants
Only the earth in every eye
Brazilians, go with us
They go shouting, but without a voice
North to south
they have no place

I can't understand that in this immense nation
it's still kill or be killed
for a piece of land

Beloved country, O Brazil
Whose are you, O gentle mother
I insist on asking

The hungry, the slums
or those who divert funds
to champagne and caviar

I can't understand
finding the clear reason
of those who only live to have
And still call themselves good Christians.

DEDICATED...

To all the people who directly and indirectly contributed to the realization of this work. If we were to list the names of all those who have helped us, this list would undoubtedly be immense and we could do ourselves the injustice of forgetting someone. So, our eternal thanks to everyone who has helped us in one way or another! Finally, thank you mom for your existence!

THANKS...

To all the migrant cane cutters interviewed for their valuable testimonies! To all the citizens of Vidigal, to whom I am proud to belong. Finally, thank you mae for your existence and my partner Joelma!

PRESENTATION

This work, which has the appearance of a book in Geography, attempts to expose the results of research carried out during my undergraduate studies and the three years of Scientific Initiation as a CNPq/PIBIC scholarship holder. We are emphasizing this character because we believe that *"a path is made by walking"* between fears, difficulties, bets and certainties. As the best summary of a text is its title, among many sleepless nights, others with many thoughts, I titled this work with a play on words that gain meaning throughout the text: "From the Dream of Freedom to Migration to Capital in the Sugar Cane Plantations of Gastâo Vidigal-SP".

The territory selected for the research was the municipality of Gastâo Vidigal-SP, located in the northwest of the state. The reasons for this choice include a mixture of objectivity and subjectivity, i.e. experience of the place, but also ignorance of the degrading situation of migrant workers in the municipality.

In general, the main objective of this research is to reveal the actions and strategies of sugarcane agribusiness, from the perspective of migrant labor in the sugarcane fields, with the aim of understanding the relations of labor exploitation, taking as a territorial basis the specific contradictions of what is happening in the municipality in question, and the spatial relations that link them to their places of origin. This research is allowing us to uncover the content of the "knots" of sugarcane migrants, why they migrate to the capital, what assessments are contained in their thoughts about this process that has been repeated and intensified in recent years.

Therefore, in order to achieve the aforementioned objectives and to produce this work, we used a methodology that combines theoretical and practical experiences. As a rule, one of the main research actions revolved around the process of bibliographical revision, in other words, throughout this journey we have tried to strengthen and deepen our theoretical reflections through bibliographical research and different readings of books, theses, dissertations, articles, magazines, publications by bodies such as: Serviço Pastoral do Migrante (SPM) and Centro de Estudos Migratórios (CEM), which deal with the theme of work and specifically migrant work in sugarcane fields.

In addition to our reading, we also dedicated our research efforts to monitoring general and specific information about labor activities in the sugarcane agro-industry, the dynamics of its expansion into the west of São Paulo and the equation of all the land set aside for this activity. To do this, we used available data from bodies such as: IEA, IBGE-SIDRA (IBGE Automatic Recovery System), MAPA, IPEA, UNICA, UDOP.

With regard to labor relations, we searched for records of interventions by the Public Ministry of Labor in the area in question. We used primary and secondary information from different sources, such as the Gastâo Vidigal-SP City Hall and the local STR, as the basis for our references to the reality of migration in the municipality.

In addition to these procedures, we are also carrying out "concrete experiences" and, in the case of geography, we are emphasizing fieldwork to recognize the local reality. In other words, we started to go out into the field and identify the social subjects and their different territorialities. To do this, we conducted "conversations/interviews" with migrant workers, especially at weekends and on public holidays when there is a greater chance of meeting them, either in their places of shelter or in the various bars in the city. However, we don't limit the right of local residents to have their say on the migration process, in other words, we interview the local population in general.

At a third point in our journey, we began to tabulate, formulate, map, analyze and interpret the primary and secondary information obtained in the previous stages, so that we could later draw up a preliminary picture of the dimensions of the research object.

The development and maturing of the research path was also enriched by the activities carried out at the Center for the Study of Labor Geography (CEGeT) and the "Florestan Fernandes" Trade Union Memory Center (CEMOSi), to which we are affiliated. We couldn't fail to mention the valuable contribution to the development of the research strategies and to the course of the research itself provided by all the discussions and colloquiums held with our advisor. Throughout this research journey, discussions and conversations with our advisor have been extremely pertinent in (re)drawing up the plans to achieve the work plan initially proposed.

Based on the data and information collected in the field, the reflections provided by the bibliographical review and discussions with the supervisor, as well as the interviews conducted with the various protagonists (migrant workers, local residents, trade unionists, the mayor and others), we have systematized this monograph into four chapters.

In order to develop this discussion, we will take the following path: initially, we will radically highlight in the Introduction the *"state of the art"* operated by the system of capital's social metabolism which, through its official discourse, sees agribusiness as the only way out for the "development" of *the* Brazilian countryside; beforehand, we will try to indicate what this "development" is, qualifying it as a development of and *for* capital. Next, in the first chapter, entitled "Appropriation and Monopolization of the Territory by the Sugarcane Agribusiness in Gastâo Vidigal - SP", we will take a brief look at the sugarcane sector, especially focusing on the real reasons for the recent expansion of sugarcane activity in Brazil, and especially in the municipality of Gastâo Vidigal-SP, highlighting the regressive impacts of this model which weakens food sovereignty and food production and obscures and silences the debate on the agrarian question behind the discourse of productivity.

In the second chapter, entitled "From the Dream of Freedom to Migration for Capital", we turn our attention to the phenomenon of migration for capital, i.e. (sugarcane agribusiness), of the northeastern migrant workforce to the municipality of Gastâo Vidigal-SP. In this section, we highlight who the migrant workers are, the reasons and places of departure and the process of recruitment by the agents, Also in this chapter, in dialogue with the readings and the testimonies of the migrant workers, we defend the assertion that migration from a socio-spatial perspective is a forced displacement through the territory, rather than a "free" decision by the social subject, in these terms, we emphasize that capitalist freedom is only a notion of bourgeois legal ideology, freedom does not really exist for the worker. The third chapter, entitled "Degradation of the Worker in the Sugarcane Plantation: Working Conditions and the Lives of Migrants", shows how the regressive work process in the sugarcane plantations is carried out and what the living conditions of the migrants are like. Then, in the fourth chapter entitled "Migration and Labor Class (Dis)belonging", we discuss issues relating to labor class (dis)belonging and territorial cleavages within migrant labor for sugarcane cutting. Finally, we make our concluding remarks defending our horizon of analysis and present the references.

INTRODUCTION

This work is the result of and is linked to a wide range of discussions on the geography of work, especially the research project "Territories in Dispute and the Geographical Dynamics of Work and Class Struggle in Brazil on the Threshold of the 21st Century", which has been developed within the scope of the Center for the Study of Geography of Work (CEGeT), under the direction and responsibility of Professor Antonio Thomaz Jûnior.

The search for knowledge and understanding of the geographic dynamics of society and, in particular, of work, has occupied my concerns and questions throughout my degree, and it is in this movement that we set out to grasp the contradictions, contents and territorial meanings of the different expressions of work, which go beyond rural and urban boundaries and the delimitations of formal and informal work relationships. This exercise is enabling us to think about the need to review theories, concepts, propositions, etc.

In this work we bring together the efforts and development of previous reflections to understand the spatial dynamics of the process of *migration for capital*, proposing to understand the spatial rearrangements promoted by the territorialization of sugarcane agribusiness, from the perspective of the migrant worker in the sugarcane fields of Gastâo Vidigal-SP. Our research assumptions are enabling us to understand the relations of labor exploitation and the impacts on the living and working conditions of migrant workers in the sugarcane fields.

First of all, we would like to emphasize the social commitment of research carried out in the academic/university environment. We don't understand it in any other way than this, because otherwise it would have no validity for the subjects who are at the mercy of capital's wheel of life on a daily basis. In this particular aspect, we emphasize the importance of a geographical understanding of the category of work. In other words, based on the construct of the societal edifice we are experiencing, we believe that the complex web of contradictory relationships that qualify the world of work at this turn of the 21st century, even with the internal fragmentation of the trade union sector, heterogenization and corporate divisions, it is possible to glimpse social emancipation from the different fronts of internal actions in the world of work that represent the new forms of social confrontation against capital (THOMAZ JR, 2007).

Guided by this horizon of analysis and the desire to transcend and reveal the hidden face of agribusiness, we tried to focus on what lies behind, and not just what our senses show us - the mere green landscape of sugar cane fields. But what is agribusiness? Historically, the notion of agribusiness was formulated in the United States by economists John H. Davis and Ray A. Goldberg (1957). For them, agribusiness involves a complex of purchasing and distributing agricultural supplies, production, storage, processing and distribution of finished products. According to Welch and Fernandes (2008), agribusiness is characterized by highly mechanized agriculture, using modern technological packages, large-scale production, exploiting large tracts of land, generally monoculture-based and geared towards export markets. This model shows a process of integration of different sets of production systems, including agricultural production, inputs, machinery and equipment, distribution, marketing, markets and financial operations. This means that the territory under the aegis of agribusiness tends towards the mono-functionality of land use and the infrastructures installed exclusively for the expanded reproduction of capital, denying the possibility of plural use of the territory.

In this way, capitalist agriculture or agribusiness, whatever euphemism is used, cannot hide what lies at its root, in its logic: *concentration and exploitation.* In this new phase of "development", agribusiness seeks to represent the image of productivity, of generating wealth for the country. As Fernandes (2008) points out, this new image of agribusiness was built to renew the image of capitalist agriculture, to "modernize" it. It is an attempt to hide the concentrating, predatory, expropriatory and exclusionary nature of agribusiness in order to emphasize only its productivist nature, highlighting the increase in production, wealth and new technologies. From slavery to the satellite-controlled combine harvester, the process of exploitation and domination is present, the concentration of land ownership intensifies and the destruction of the peasantry increases, thousands of men and women are forced to move around in search of survival, while the latifundium carries out exclusion through unproductivity, agribusiness promotes exclusion through intense "destructive productivity".

Thus, we can say that the official discourse of capital constructs a sense of modernity and wealth, as if these were shared equally by all, distributed homogeneously and generating collective well-being. In this sense, agribusiness is seen as the only form of agricultural development, all the others would be remnants, archaic forms on the verge of disappearing, only agribusiness would lead to "development", in these terms we can see an intimate discursive relationship between agribusiness and development.

> The ideologization of the low productivity of the land, or the economic "powerlessness" of peasants/producers and land workers to continue producing food (indebtedness, difficulty in accessing credit, added to agricultural policies that are unfavourable to them), supports the expansionist project of sugarcane agro-industrial capital and alters the social relations of production and work in the countryside, thus revealing the new faces of the contradiction inherent in the metabolic system of capital, personified in sugarcane agribusiness, which claims to be the bearer of progress and development, thus corroborating the hypothesis that a geographic (re)configuration is taking place in the territories where this project materializes (OLIVEIRA, 2009, p. 333). 337).

Thus, agribusiness assumes the role of protagonist of "development", so some questions arise: What is development? What kind of development are we talking about? I'll try to develop this reasoning, but the answer seems logical to me: it's actually a development that takes place within the assumptions of capitalist socio-spatial formation, it's a *development for capital.*

It can be said that the discourse of development has been consolidated, especially since the second half of the 20th century, based on the ideological construct that modernization overcomes backward relations and forms of production; that industrialization and technology lead to modernization; and that the materialization of this development guarantees progress in the social, cultural and political spheres. This supposed "development" is actually the contradictory logic of capital, which in its process of production and reproduction, manifests its unequal and combined essence in the territory. In other words, development, contrary to what is proposed, is also contradictory and, when it materializes in geographic space, it reveals the inequalities generated by the contradiction. In this sense, Montenegro Gómez points out that the development promoted in capitalist society "pretends" to reduce inequalities through a system based on competition between unequals. At the very least, this is a nonsense, if not a cruel imposture (2006, p. 121).

In this way, the "development" postulated by agribusiness is nothing more than a strategy for the reproduction of capital and social control, which, according to the official discourse, would serve to repair the temporal upheavals caused by the natural dynamics of the system, allowing the poor (underdeveloped) to enjoy, in the very near future, the benefits that the rich (developed) already enjoy. In this way, development policies would be the system's concern and conscience, coming to the aid of those most in need. In the last decade, a spectrum has been created around the idea of development based on *"social, environmental and human aspects"*, so that *"the so-called 'capitalism with a human face' is nothing more than a new face"* (MONTENEGRO GÓMEZ, 2004, p. 55).

This is because we can clearly see the contradictions and pitfalls of development:

> [...] instead of the kingdom of abundance promised by the theoreticians and politicians of the 1950s, the discourse and strategy of development produced the opposite: massive misery and underdevelopment, unspeakable exploitation and oppression (ESCOBAR *apud* MONTENEGRO GÓMEZ, 1998 [1996], p. 21).

As we want to demonstrate, agribusiness is based on the discourse of development and the generation of jobs and income for workers. However, its concrete developments do not aim to improve the lives of the poorest populations, and this is not even remotely its objective. Instead of being the "salvation of Brazilian agriculture", this business-society and development model is completely bankrupt in terms of its impact on workers.

It is therefore a development of capital and for capital, which is based on the dismantling of communities and territories that are of interest to capital's expansionist project with a view to their valorization, taking away their own historically constructed conditions and imposing actions and attitudes that do not match previous forms of social organization and production (OLIVEIRA, 2009, p. 344-45).

In other words, the idea of development, within the *commodity-producing society* in which we live, brings us back to the structural subordination of labour to capital, so adjectives like human, solidarity or sustainable, associated with local development, are attempts to harmonize the destructive logic of capital in relation to the human race. An attempt doomed to failure. In this respect, it is worth noting that despite the country's agricultural potential, millions of people do not have access to basic food. According to data from IBGE-PNAD (2013), 7.2 million Brazilians are hungry and more than 52 million live in a situation of "food insecurity". In other words, this is a development model that, instead of guaranteeing access to land for millions of workers, through Agrarian Reform, and prioritizing Food Sovereignty, legalizes land grabbing, increases land concentration and accentuates the exploitation of nature and the human workforce.

CHAPTER I . APPROPRIATION AND MONOPOLIZATION OF TERRITORY BY SUGARCANE AGRIBUSINESS IN GASTAO VIDIGAL- SP

The current process of sugarcane expansion in Brazil is backed by the discourse of agribusiness, which is being imposed as an efficient economic development model that generates jobs and income and tries to mask the image of capitalist agriculture, which has historically exploited labor and concentrated land, income and capital. What stands out in this process is the dynamic of the territorialization of capital in the countryside, based on the production of an (ufanistic) discourse on agribusiness as the only development model capable of solving the country's economic problems. In other words, the consolidation of a discourse based on a single development model, the development of agribusiness as the "salvation of agriculture".

The characteristics that marked and still mark the sugarcane agro-industry in Brazil are important clues for understanding the challenges imposed on work in the 21st century. To enter this debate, we can point out, as Souza (2008) does, that the sugarcane agro-industry is the oldest non-extractive economic activity in Brazil. Its main product, from the middle of the 16th century (1540/50) to the end of the 20th century (1979/80), was sugar. Ethyl alcohol took over for a short time. After the reversal of the upward trend in international oil prices from 1983/84 and after a few years of difficulties, sugar regained its primacy in the 1990s, with a lot of help from exports.

We all know that the establishment of the sugarcane sector in our country, since colonial times, has been based on latifundia, slave labor, the monoculture of sugarcane for export and the mistreatment of workers. Taking a long historical leap, we can say that this concentrationist nature of land, income and capital in the sector was strengthened with the creation of the SNCR (National Rural Credit System), set up by the military government in 1965.

According to Oliveira (2007), by creating a highly selective and exclusionary policy of granting subsidized credit, the state ended up favoring large landowners and business groups, contributing to the creation of the Brazilian Agro-Industrial Complex and, in particular, the sugarcane CAI. For the author, the changes instituted in the sector from then on, with the aim of increasing the scale of production and raising the organic composition of capital, were based on land speculation and the intensification of the concentration of land ownership, resulting in the expropriation of small landowners, settlers and squatters and, consequently, an increase in wage labor in the countryside.

In the 1970s, from 1975 onwards, the federal government created Proàlcool (National Alcohol Program), which had as "one of its objectives" to increase the production of fuel alcohol due to the oil crises on a global scale that were seriously affecting the balance of payments; For as Thomaz Junior (1992) points out, Pro-Alcohol, above all suspicion, before representing an energy alternative, was the economic way out that the state once again provided for the sector (at a level not comparable until then), through subsidized financing and credit facilities, at the expense of the production of food for the basic food basket.

In general terms, according to the Ministry of Industry and Trade (1986)[1] , Pro-Alcohol in its first phase took advantage of idle capacity by setting up distilleries attached to sugar mills, so alcohol production went from 600 million liters in the 1975/76 harvest to 3.6 billion liters in the 1978/79 harvest, destined only for the gasoline/anhydrous alcohol mix.

The second oil shock in 1979 allowed Pro-Alcohol to enter its second phase, for the production of hydrated alcohol. In this second phase, emphasis was placed on setting up autonomous distilleries in new production areas. And finally, from 1985 onwards, Pro-Alcohol entered a new phase, having to live with the reduction in the volume of money, with the closing of the tap.

1 Cf. BRAZIL. Ministry of Industry and Commerce, 1986.

What are the consequences of this program?

It's worth noting that one of the direct consequences of Pro-Alcohol was the vertiginous increase in the production of sugar cane and, consequently, its by-products, and that it was extremely profitable for agro-industrial capital, fulfilling all the wishes of the sugar cane sector, including centuries-old debt forgiveness .[2]

According to Oliveira (2007), the intense transfer of income from the social capital as a whole to the sugarcane sector, especially since Pro-Alcohol, was capitalized differentially by the sugarcane agro-industrial companies, making it possible to concentrate capital in the hands of a few groups/families.

In the 1990s, the state, unable to continue with its credit policy due to the fiscal crisis and also due to the change in the paradigm of economic policy, linked to the neoliberal assumptions of the Washington Consensus, promoted the deregulation of the sugarcane agro-industry, or more specifically the sugarcane sector, and the opening up of trade.

Thus, Alves (2007) points out that these two directions, opening up trade and deregulation, operated at the same time and combined with the fall in international sugar prices and the fall in domestic demand for ethanol, led to a crisis in the sector in the 90s. The withdrawal of the state and the opening up of the domestic market led to significant changes within the complex, as Alves (2003) points out below:

> Internally, the companies in the sector have started to compete more fiercely with each other than they did before, when the state administered prices for (cane, sugar and alcohol). Whereas before competition between mills was only for the supply of cane, or for land, or even for special favors from the government, today competition is internalized between mills (p.78).

In other words, the partial withdrawal of the state transferred competition to the sugarcane agro-industrial sector, and in this new scenario a process of productive restructuring developed in the sector with the introduction of a set

2 Cf. THOMAZ JR, 2002.

of modifications ranging from changes in the technical base of production to organizational changes in production and work processes.[3]

Recently, the sugarcane agro-industry has been experiencing a new moment of expansion. According to Alves (2008), from 2002 onwards the sector entered a period of great dynamism and growth, reminiscent of the golden age of Pro-Alcohol (1975/1983), except for one essential difference: the absence of direct and exclusive subsidies guaranteed by the state.

This current dynamism is due to the privileged role that its two main *commodities* - sugar and, above all, alcohol - enjoy in world agribusiness. So recently, the rise in oil prices on the international market, the environmental problem and the adoption of public policies by developed countries to encourage the use of "alternative" and less polluting fuels have justified the *boom* in Brazilian sugarcane and ethanol production. Alongside the United States, Brazil occupies the leading position in the world sugar-alcohol sector and is attracting a wave of foreign investors to the country. It's worth remembering that the US has quickly become the world's largest producer of ethanol, but uses corn to produce it.

It is in this context that the political and economic situation of the opening up of the ethanol market has also contributed to the emergence of territorial rearrangements, both in terms of capital and labor. The great driving force behind this recent process of expansion in the sector is the possibility of conquering foreign markets and making ethanol a *commodity*.

So much so that Oliveira (2007) assures us that since the 2000s, the geography of sugarcane in Brazil has been changing significantly. It is worth noting that in the state of São Paulo, this recent expansion is occurring more intensely in the west, especially in the administrative regions of Araçatuba, São José do Rio Preto and Presidente Prudente. In other words, this movement of capital within the sugarcane sector has been expressing a new territoriality, or

3 Cf. ALVES, F., 2008.

why not say, a new territorial planning of agro-industries and areas planted with sugarcane. If, until the end of the 1990s, sugarcane activity was concentrated in the main producing states in the Northeast (Alagoas, Paraiba and Pernambuco) and in the state of São Paulo (the country's largest producer), what we are seeing today is a growing movement of expansion into other Brazilian regions and states which, until then, had not had any expression in national sugarcane production. With regard to the sugarcane phenomenon, the process is expressed as the spatialization of agro-processing plants and plantations, based on a recent process of expansion never seen before.

As Oliveira (2009) points out, the area harvested in Brazil with sugarcane in the 2009/2010 harvest, based on data from the National Supply Company (CONAB, 2010), was estimated at 7.531 million hectares, distributed across all the producing states, with the greatest concentration in São Paulo - with 4,101.4 million ha; followed by Paranà - 590.1 thousand ha; Minas Gerais - 587.1 thousand ha; Goiàs - 520.3 thousand ha and Alagoas - with 448 thousand ha. In terms of the quantity of sugar cane produced, in 2009 production reached 612.2 million tons, up 7.1% on the previous year. The state of São Paulo accounted for 57.8% of production and harvested 354.3 million tons, 2.5% more than in 2008. Three other states stood out in terms of the increase in the amount of cane harvested. These were Goiás, with production up 50.1% and reaching 44.4 million tons; Mato Grosso do Sul, with an increase of 38.8%, accounting for 28.8 million tons; and, finally, Minas Gerais, where the increase was 20.1%, with 49.7 million tons harvested.

This growth, driven by the explosion in the sale of "*flex-fuel*" cars in Brazil, is being made possible by the territorialization and expansion of sugarcane agro-industrial capital into non-traditional areas for growing grass, and it is at this moment that the west of São Paulo is attractively inserted into the unbridled advance of monoculture, with enormous possibilities for mechanization given the relief conditions, and with a significant stock of land. In the western region of the state of São Paulo, traditionally used for beef cattle, the area given over to cultivated pasture between 2001 and 2006 corresponded to 62% of all the area given over to this activity in the state of São Paulo. Among the crops that gave up the most area during this period were corn and basic food crops such as beans, with a reduction of around 55% in production, and rainfed rice, with a 61% reduction in production (DANIEL et al, 2009).

And this appropriation and monopolization of the territory by agribusiness has been expressed in the dynamics of the *disputed territories,* in the territorial section of our approach, the municipality of Gastâo Vidigal-SP, part of the 9ª Administrative Region of Araçatuba and which is part of the EDR (Rural Development Office) of General Salgado-SP (Table 1), forming the following territorial base (Map 1).

Table 1. Municipalities in the EDR of General Salgado - SP

Escritório de Desenvolvimento Rural de General Salgado-SP Municípios Integrantes
1- Auriflama
2- Buritama
3- Floreal
4- Gastão Vidigal
5- General Salgado
6- Guzolândia
7- Lourdes
8- Macaubal
9- Magda
10- Monções
11- Nhandeara
12- Nova Castilho
13- Nova Luzitânia
14- Planalto
15- Santo Antônio do Aracangua
16- São João de Iracema
17- Sebastianóplis do Sul
18- Sud Mennucci
19- Turiúba
20- União Paulista
21- Zacarias

Source: Gastâo Vidigal City Hall - SP

Map 1- Location of the Municipality of Gastão Vidigal- SP

Source: Cardoso, 2010.

According to data from the IEA - Instituto de Economia Agricola (2009), sugar cane production in the EDR of General Salgado increased from 3,924,362t in 2000 to 11,443,939t in 2008, which corresponds to an increase of 192%, much higher than the increase in the state of São Paulo over the same period and amounting to approximately 3% of the total produced in the state.

In this context, of the expansion of sugarcane production on a national scale, we can point out that this concomitant trend also applies to this territory in the General Salgado-SP EDR, which is part of the recent expansion of the Agrohydronegócio[4] sugarcane polygon, as Graph 1 shows.

Graph 1- Territorialization of sugarcane production (tons) in the EDR of General Salgado-SP

4 Name borrowed from Thomaz Junior, 2009.

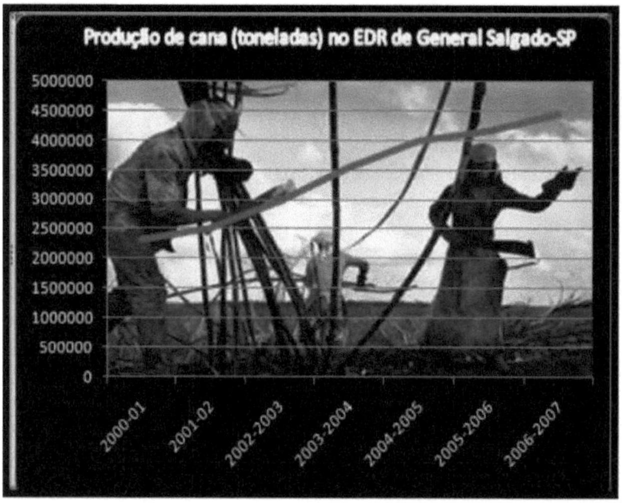

Source: UDOP, 2011.

Based on the data above, we can say that the territorialization of sugarcane agribusiness is taking place in the General Salgado-SP EDR, and that this process gained momentum on the threshold of the 2000s. In this sense, the growth in sugarcane production is notorious. In detail, from the 2000/2001 harvest to the 2006/2007 harvest, there was an increase in sugarcane production of around 90.3%, in other words, we can characterize our territorial section of the study as being a "territory in dispute", in which it is in the interest of capital to deepen the process of its insertion into the sugarcane circuit. In this way, the expansion of sugarcane monoculture in the EDR of General Salgado-SP can be seen in Maps 2 and 3 and dimensioned in Table 2.

Map 2- Temporalization of sugarcane plantations in the EDR of General Salgado-SP Crop 2003/2004

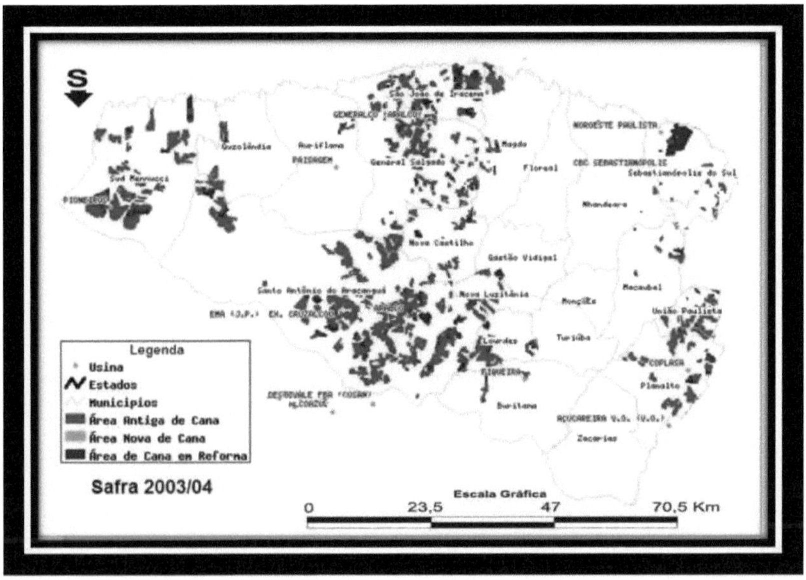

Source: Canasat, 2010/Cardoso, 2011.

Map 3- Territorialization of sugarcane plantations in the EDR of General Salgado-SP Crop 2008/2009

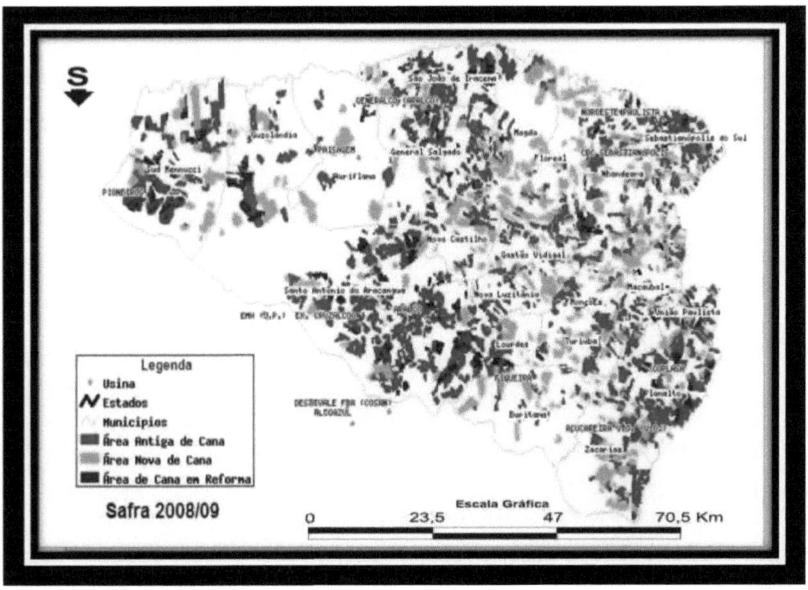

Source: Canasat, 2010/Cardoso, 2011.

Table 2. Sugarcane area per municipality in the General Salgado-SP EDR in the years 2005 to 2008, and their respective growth variation over this period.

Municípios	Área (ha)				Variação (%)
	2005	2006	2007	2008	2005-2008
Auriflama	778	280	2.180	2.180	180,2
Buritama	470	1.070	3.000	3.000	538,3
Floreal	157	1.175	4.100	4.600	2824,3
Gastão Vidigal	230	2.527	5.247	5.247	2181,5
General Salgado	6.775	7.475	8.605	9.640	42,3
Guzolândia	955	1.977	3.218	3.377	253,5
Lourdes	1.507	1.604	12.731	4.330	187,3
Macaubal	1.500	1.540	2.500	2.400	60,0
Magda	4.300	6.000	8.000	8.000	86,0
Monções	720	2.160	2.160	6.000	733,3
Nhandeara	6.200	6.600	10.500	15.200	145,2
Nova Castilho	2.000	3.500	5.300	4.500	125,0
Nova Lusitânia	1.258	1.420	1.948	1.780	41,5
Planalto	6.000	9.000	13.000	13.000	116,7
Santo Antonio do Aracanguá	19.692	22.010	22.010	22.010	11,8
São João de Iracema	4.800	4.800	5.800	7.600	58,3
Sebastianópolis do Sul	3.400	6.000	5.000	5.500	61,8
Sud Mennucci	8.900	9.673	9.550	12.650	42,1
Turiúba	450	1.874	5.100	5.100	1033,3
União Paulista	1.691	1.740	4.558	4.520	167,3
Zacarias	2.700	5.000	7.700	8.800	225,9
EDR de General Salgado	**74.484**	**97.425**	**142.208**	**149.435**	**100,6**
Estado de São Paulo	**3.673.275**	**4.258.370**	**4.835.434**	**5.389.303**	**46,7**

Source: IEA basic data, 2009.

By reading the maps and looking at the table, it is possible to confirm the idea that on the regional scale of the General Salgado-SP EDR, to which the municipality under study belongs, the process of land grabbing by the sugarcane agribusiness has been increasing and controlling an ever larger portion of the territory, As the table above shows, the expansion of sugarcane agribusiness from 2005 to 2008 reached 100.6%, well above the state percentage of 46.7%. But to what extent does this dynamic of expansion affect the distribution and destination of agricultural uses, especially in Gastâo Vidigal - SP? In this context of the expansion of sugarcane activity, we can highlight that this macro-scale trend is reflected at the micro-scale level in the municipality of Gastâo Vidigal, with land being taken over to enter the sugarcane circuit (Graph 2).

Graph 2. Territorialization of sugarcane agribusiness - Land appropriation for sugarcane plantations in Gastão Vidigal-SP

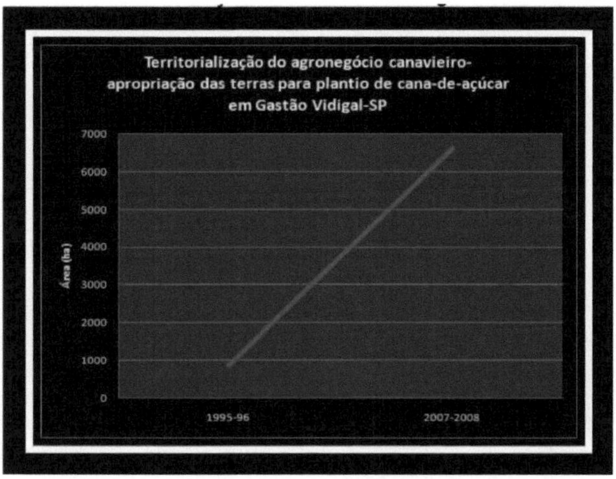

Source: Cati- Lupa, 2011 (Census Survey of Agricultural Production Units in the state of São Paulo)

The sugarcane plantations are advancing intensively on the lands of the municipality of Gastâo Vidigal-SP, and according to the data, the area planted with sugarcane in hectares, from 1995-96 to 2007-08, grew by approximately 771%, reaching the exorbitant appropriation of 6625 thousand ha of land, while the process is reversed when it comes to the lands destined for pasture (Graph 3) and mainly for food production (Graphs 4 and 5).

Graph 3. Reduction in pasture area in Gastâo Vidigal-SP

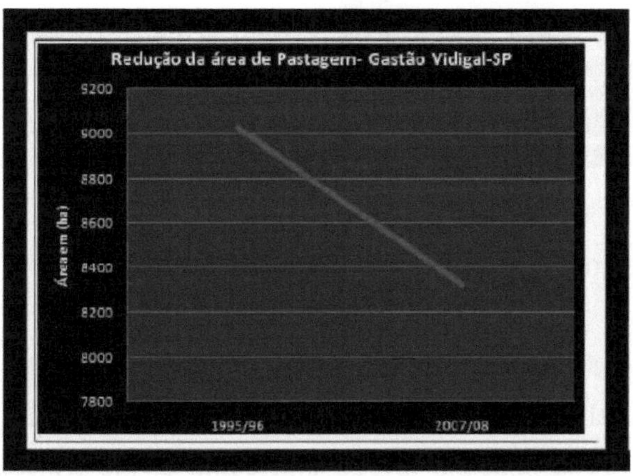

Source: Cati- Lupa, 2011 (Census Survey of Agricultural Production Units in the state of São Paulo).

Graph 4. Reduction in the area set aside for corn production - Gastão Vidigal-SP

Source: SIDRA/IBGE - Municipal Agricultural Production, 2011.

Graph 5. Reduction in the area set aside for rice production (in husk) - Gastâo Vidigal-SP

Source: SIDRA/IBGE - Municipal Agricultural Production, 2011.

As for the reduction in pastureland, the pace is slow. In percentage terms, from 1995/96 to 2007/08, there was a reduction of approximately 7.74%, which is nothing compared to the reduction in areas used for food production. According to data on municipal agricultural production from the IBGE (Brazilian Institute of Geography and Statistics), it is possible to see that some of the typical products of peasant food agriculture in Gastâo Vidigal-SP, such as corn and rice, showed figures during the period analyzed, Just as an example, the area set aside for corn production fell by approximately 574% from 2003 to 2009, while in the case of land set aside for rice production the reduction was total, i.e. the basic food crop of rice became extinct in the municipality. What we want to highlight is that while the data on land available for traditional food crops shows a drop, a sharp decline in the area planted, the opposite is true for sugarcane agribusiness. As such, we would like to defend the idea that the growing appropriation of land by agribusiness in the municipality of Gastâo Vidigal-SP has also been taking place on areas of food crops (Graph 6). Therefore, the local dynamics of the insertion of land into the sugarcane circuit has demystified the studies pointed out by the "mainstream press", which claim that food production will not be affected by the expansion of the "sugarcane sea".

Graph 6. Reduction in areas used for food production-Gastâo Vidigal-SP-1995-96/ 2007-08.

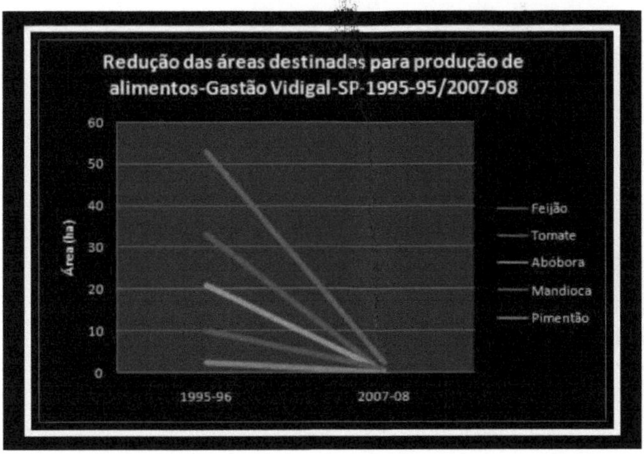

Source: Cati- Lupa, 2011 (Census Survey of Agricultural Production Units in the state of São Paulo)

As we can see from the graph above, basic food crops for humanity in the municipality under analysis are on the verge of disappearing. Just as an example, the area set aside for the production of beans, a crop that used to be very important in the municipality, has had its land reduced by 2650%, currently covering an insignificant 2.4 ha, while others such as tomatoes and peppers have already been extinguished from municipal production. In these terms, the model proposed by sugarcane agribusiness does not meet food sovereignty, quite the opposite, it attacks peasant food production, in other words, it would be the same as saying that the appropriation of territory for sugarcane production under the agribusiness development model is incompatible, incongruous with food production. It is clear, then, that the capitalist appropriation of territory promoted by the advance of sugarcane agribusiness on a local scale tends towards the homogenization of the landscape and the monopoly and control of land use and occupation (Graphs 7 and 8).

Graph 7. Land use and occupation 2007/2008 in Gastâo Vidigal-SP

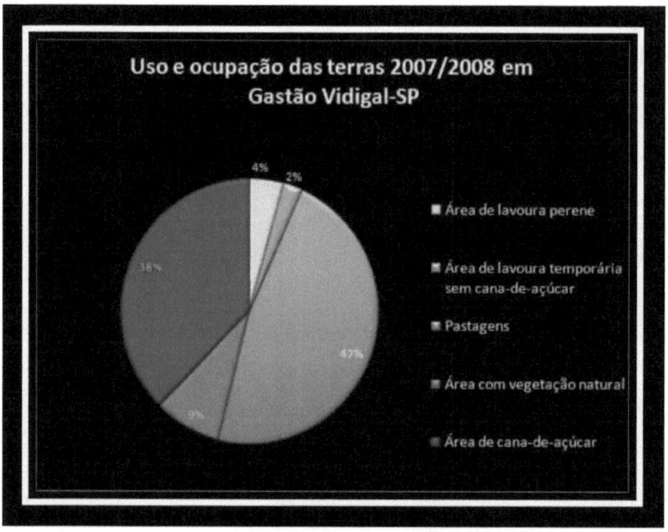

Source: Cati- Lupa, 2011 (Census Survey of Agricultural Production Units in the state of São Paulo)

Graph 8- Use and occupation of agricultural land 2007/2008 in Gastâo Vidigal-SP

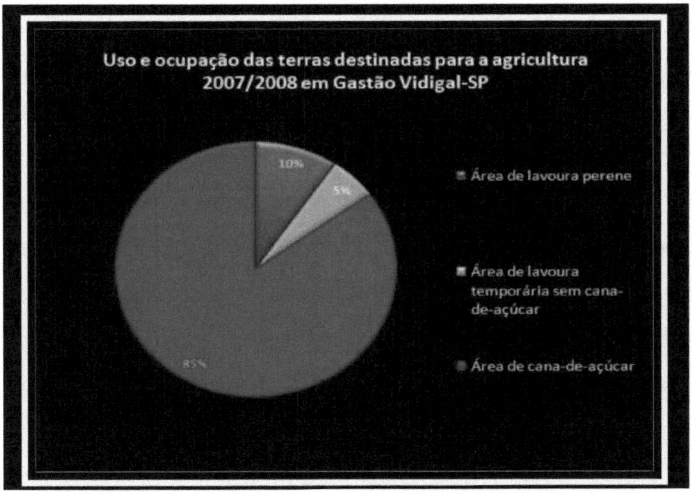

Source: Cati- Lupa, 2011 (Census Survey of Agricultural Production Units in the state of São Paulo)

For those less aware, this is not just a *harmless territorial appropriation*, but a *destructive territorial appropriation* that attacks relationships that don't fit into its game. This is how we can interpret the graphs above, which show that the territorial portion already occupied by sugarcane agribusiness is in the order of 38% of the entire municipal territory and the tendency is for it to advance intensely over the rest of the territory, Furthermore, sugar cane already occupies 85% of the land set aside for agriculture alone, thus confirming the idea we have been defending that the capitalist appropriation of the municipal territory by sugar cane agribusiness *is taking off in monocultures* worthy of the colonial period, so agro-industrial capital already has territorial control of more than 1/3 of the municipality of Gastâo Vidigal-SP.

According to data collected from CATI/Casa da Agricultura in the municipality, around 45% of local agricultural production units are part of the sugarcane production circuit. How can we explain such territorial control? According to Backes (2008), the process of "homogenizing" the territory through the sugarcane agro-industry is slowly affecting small farmers, tenants and salaried workers in the countryside, appropriating and expanding throughout the territory. Going further, it is interesting to highlight the participation of land as a commodity in this process. As (MARTINS, 1983, p. 159-160) states.

> Land is not the product of wage labor or any other form of labor. It is a natural asset that cannot be reproduced, it cannot be created by labour (...), land is not the materialization of human labour, but is transformed into merchandise.

As Ianni (1981, p.154) points out, "land becomes a commodity, it gains in price, it becomes an object and a means of producing exchange values, and it is part of capitalist production relations". As a general rule, in terms of the expansion of sugarcane plantations in Gastâo Vidigal-SP, we would like to defend the assertion that the great motivator when making the decision to lease land to capital lies in the economic factor which, at the beginning of the process, due to the difficulty in obtaining credit and rural assistance for traditional agricultural cultivation, ends up being "attracted" by the high and illusory price paid in the moments of euphoria of the sugarcane sector.

When the land is leased back to the mill owner at the end of the lease period, when there are no resources to re-establish the activities previously practiced, when it is impossible to rebuild fences, pastures, houses and all the infrastructure removed by the formation of large areas of monoculture plantations, plus the fact that agricultural machinery and implements are scrapped and/or sold because they are no longer used, that's when the "*cat jumps*", where the owner finds himself hostage to capital.

In this way, weakened by the unequal relationship with big capital, the farmer/owner ends up with two equally unfavorable alternatives: either he renews the lease for a much lower price (the mills' payment is proportional to the production of the planted area), because in a second planting opportunity the production will probably be lower, due to the wear and tear on the soil. The other path leads the landowner to give in once and for all to the entourage of sugarcane capital, selling his land for much less than it was worth before the lease, becoming a proletarianized workforce for the expansion of sugarcane production, giving materiality to the expansionist/monopolizing/territorializing intent of agribusiness.

At the end of this chapter, based on our research, I propose that the apology for agribusiness by the media, companies and the state is a way of shielding this model, trying to make its immanent conflict invisible.

> The development of agriculture also takes place through the perennial conflict between social classes, who clash and dispute over politics and territory. This *development through conflict* is characterized by insoluble contradictions and paradoxes, which have been recorded in modern history predominantly through the territorialization of capital and the deterritorialization and reterritorialization of the peasantry (FERNANDES, 2004, p. 9).

In this way, agribusiness tries to silence its impacts by positioning itself as a productive space par excellence, whose supremacy cannot be threatened. Agribusiness, unless I'm mistaken in my comparisons, is a new type of latifundia[5] and an even broader one. Now it not only concentrates and dominates land, but also production technology and "development" policies.

In this sense, this dynamic represents the materialization of all the ideals of public policies that value agribusiness to the detriment of peasant-based agriculture linked to food production. Thus, this change in the productive structure of this municipality reveals the meaning and magnitude of the territorialization of capital in the countryside, in other words, we are faced with a composition of forces that monopolizes the use of and ensures access to land for capital. Food production, security

the capitalist relations established in the countryside, because what is on the agenda is the exploitation of the land based on monoculture and the emphasis on export-oriented production.

In this way, the landscapes and the components that give meaning to the territory are destructively changed for the expanded reproduction of capital. Thus, the capitalist appropriation of the territory takes place through the homogenization of sugar cane monoculture, which establishes a "green desert" as capital's landscape, imprisoning the landscape and depriving it of its

---

5 Cf. FERNANDES, 2004.

meaning for humanity. This problem is created by the logic of the expanded reproduction of capital, which causes uneven development through the concentration of power expressed in different forms, for example: ownership of land, money and technology. This logic produces the concentration of power by creating the power to concentrate, reproducing itself infinitely. Endless reproduction is in the nature of the capitalist mode of production, so in order to guarantee its existence, capital needs to territorialize itself without limits. And in its *eagerness to territorialize*, capital tries to destroy and advance on other territories.

CHAPTER II. FROM THE DREAM OF FREEDOM TO MIGRATION TO CAPITAL

In this chapter, we will try to understand the contradictions in the process of *migration from labor to capital,* in the case of migrant workers cutting sugar cane in the municipality of Gastâo Vidigal-SP. The theoretical positions linked to the concrete experiences gained through fieldwork and contact with the workers have made the difference in developing the ideas that will be presented here.

The spatial mobility of labor has been a recurring theme in various studies in the fields of geography, economics and other social sciences, with different theoretical approaches. For the purposes of this chapter, we distinguish between two antagonistic currents to which scholars of this phenomenon have subscribed: on the one hand, neoclassical economics and, on the other, dialectical historical materialism.

For the first current, the spatial mobility of labor would be guided by the laws of the market and the premises of rationality and the sovereign freedom of men - leading to a spatial differentiation that would, at a certain point, find a harmonious state of equilibrium. It would only be up to the workers thus considered to find their place in the labor market, based on the demand for workers in their own or other municipalities, in their own or other regions, according to their professional possibilities. Gomes (2009) gives a similar opinion on this neoclassical current, which he calls the "general equilibrium current", when he states that:

> This trend presupposes the existence of perfect labor mobility, arguing that workers have the ability to move between spaces in search of better living conditions, which translate into higher wages and incomes. Thus, freedom of choice is attributed to the worker. The neoclassicals admit that individuals behave rationally and can choose between employment alternatives and the best material living conditions (GOMES, 2009, p. 38).

According to Vale (2007), in the neoclassical view, migration studies presuppose a rational calculation and the free decision of individuals, i.e. the free movement of bodies in space, and this movement leads to spatial heterogeneity and fosters equilibrium, in the sense of possible equality.

The second current, based on dialectical historical materialism, is totally opposed to the first. In short, for scholars committed to this perspective, capitalist social formation, by generating surplus workers, separated from their means of existence, creates the need to move around in search of work. The availability of such workers, as a market reserve, in the most diverse parts of the territory, becomes, in turn, a necessary condition for the very existence of capital accumulation. The spatial mobility of labor is not merely a mechanism for the spatial redistribution of populations, which adapts to the demands of the economic system, but is configured as structural for the accumulation of capital. For Vale (2007, p.46-7), in capitalist structures marked by the rigidity or expansion of their economic morphology, forms of labor mobility appear as a phenomenon of submission and not of freedom. For Gaudemar (1977, p. 172), "every capitalist strategy of mobility is also a strategy of forced mobility". Perpetua (2010, p. 9) summarizes this theoretical current well:

> [...] it is based, on the one hand, not on the perfect mobility of workers over space, but on forced, awkward mobility, an elementary condition of capital accumulation; and, on the other, not on the general equilibrium of spatial development under capitalism, but on the constant imbalance and contradiction, expressed by the inequality that ultimately generates the need for mobility.

As such, we reject and diametrically disagree with the neoclassical current's analysis of the spatial mobility of labor, and in this work our efforts have converged on trying to develop some reflections on one of the faces of the phenomenon of the spatial mobility of labor that has also been gradually intensifying: the migration of seasonal workers from the Northeast to cut sugar cane in the municipality of Gastâo Vidigal-SP. In order to understand this issue, we defend the idea that this process of labor migration, given the understanding that capital is hegemonic and exercises control over the entire social fabric, is not for nothing that labor migration is a *migration for capital,* that is, *a forced migration* driven by the need to survive and not a mere act of exercising "freedom".

Having made these reservations and returning to our central reasoning, we can say that since the emergence of man thousands of years ago on the African continent, the search for better living conditions has always been one of the goals to be achieved. For this reason, the first societies were nomadic, as they always migrated in search of what they had run out of. The movement of populations in space, in other words, migration, is motivated by various factors: political, religious, natural, cultural, but without a shadow of a doubt, the factor that has historically been predominant in capitalist socio-spatial formation is the economic one, in other words, migratory dynamics are structurally linked to the functioning of the capitalist mode of production which, in order to generate more value, requires the labor of human beings and the engendering of a spatial ordering of the labor force, and, if necessary, its territorial reordering.

We must understand that these movements are driven above all by capital, and that they are a way of adjusting the workforce to the needs of capital in the process of valorization (OLIVEIRA, 2007). In this sense, the migratory movement in Brazil was and is linked to important economic cycles (such as rubber, coffee, the period of consolidation of the industrial sector, etc.), and even to the process of modernization/mechanization of agriculture, which caused a significant outflow of rural people to urban centers, either because they were expropriated from their land or because their labour was replaced by machinery.

Bringing the discussion to the present day, what is pertinent to us at this point is to point out that the *geography of sugarcane* is changing in our country and, above all, it is causing decisive changes in the *geography of work*. According to Oliveira (2007), in this context in which sugarcane agro-industrial capital is (partially) deterritorializing itself and (re)territorializing itself in new places, this same process is also leading to the deterritorialization and (re)territorialization of workers, in other words, the mobility of labour is accompanying the movement of capital itself. Therefore, we would like to venture the hypothesis that the sugarcane agro-industry, in yet another of its phases of expansion, has been making use of the intense search for and exploitation of migrant labor to achieve its high levels of profitability.

The concept of deterritorialization dealt with here takes into account the fact that the human force that works lives a constant territorial mobility driven and conditioned by the self-expansionist and deterritorializing dynamics of capital, especially in the current context of globalization and the productive restructuring of capital.

As such, it is a forced deterritorialization due to the need for men, deprived of the means of production, to sell their commodity labour power to capital. Dependent on employment, wages and housing, and therefore on basic conditions to guarantee the reproduction of his life and family, the worker is led *"in the extreme to get rid of everything that had territorialized him"* (MOREIRA, 2008, p. 137), migrating to often distant places where he will have to form new social relationships, new cultural ties and a new identity, thus establishing a new territoriality (OLIVEIRA, 2009).

In general, sugarcane cutters are migrants from other regions of the country. However, as Oliveira (2007) states, a few years ago, most of the migrant labor in sugarcane cutting came from the north of Minas Gerais (the Jequitinhonha Valley) and from some states in the northeast (Pernambuco, Paraiba and Bahia), basically heading for the state of São Paulo. Today, it is possible to notice an increase in the number of migrant workers, not only to São Paulo, but to all the states of the Center South, which are expanding sugarcane production. This expansion process is even attracting workers from other states in the Northeast, such as Piaui and Maranhão .[6]

As Silva (2008) shows us, while in the Center-South sugar cane plantations

6 Other studies have also highlighted this, such as Alves, 2007; Silva, 2008; Oliveira, 2009; Thomaz Junior, 2009.

are spreading and pasture areas are shrinking, by way of example, by 15% between 1995 and 2006, in the North of the country there is a huge advance in cattle breeding in the Legal Amazon, with the number of cattle rising from 34.7 million to over 73 million between 1994 and 2006.

We can venture to say that this process has been responsible for the conversion of peasants into migrants to work in sugarcane cutting in the Southeast and Center-West or to work in the devastation of the Amazon to plant grass for the development of cattle ranching. We can summarize this process as Silva (2008, p.08) does: "Geographically, it can be said that there is a road with two opposite directions: while men go down (from the North and Northeast to the Southeast), capital (oxen go up from the Southeast to the North)."

In a categorical way, we can say that the municipality of Gastâo Vidigal-SP has been undergoing profound territorial rearrangements, as a rule, brought about by the recent expansion of the sugarcane industry, in other words, it is in the process of being actively inserted into the agribusiness circuit. According to the information obtained in the field, in terms of migrant labor, the municipality, which has 4193 inhabitants, received around 1900 migrants in 2011, according to local estimates, to cut sugar cane, and this workforce is basically made up of people from Piauí .[7]

The research process and the forays into the field allowed us to reveal the territorialization of this migratory flow, or rather, to understand where these migrant workers come from and what they are trying to achieve on São Paulo soil? In order to understand the contradictions of this process of *migrating to the capital*, we carried out 56 interviews/conversations with different migrant "cane cutters" in the municipality in question over the course of 2011 (appendix 1). Through direct contact with the migrant reality, we were able to advance in the research process and propose some reflections, (table 3) below attempts

---

7 Data collected from the municipality's STR.

to detail the territoriality of the migratory flow that converges on the municipality of Gastâo Vidigal-SP.

Table 3- Territoriality of the Migrant Flow (Piaui- Gastâo Vidigal-SP), 2011

Locations	No. of interviewees	Percentage
Esperantina-PI	13	23%
Altos-PI	10	18%
Campo Maior-PI	7	12, 5%
Piracuruca-PI	7	12, 5%
José de Freitas-PI	6	11%
Uniao-PI	5	9%
Teresina-PI	5	9%
Picos-PI	3	5%
Total	56	100%

Source: Fieldwork/Cardoso, 2011.

In general terms, the main municipalities supplying migrant labor to the sugarcane capital in Gastâo Vidigal-SP are: Esperantina-PI, Altos-PI, Campo Maior-PI, Piracuruca-PI, José de Freitas-PI, Uniâo-PI, Teresina-PI, and Picos-PI, i.e. there is a massive predominance of migrants from the state of Piauí, from the north/northwestern part of the state, characterized by small family properties destined for agricultural activity according to the 2003 Agrarian Reform Plan for the state of Piauí. Map 4 below shows the spatial location of the current flow of migrants towards the municipality. As we can see from Table 3, the municipality of Esperantina-PI is the biggest contributor of labor, although six other municipalities in the north/north-west of the state also join the wave of migrant workers heading to the sugarcane fields, only the municipality of Picos-PI is located outside this region in the east.

Map 4- Spatialization of the Flow of Migrant Workers for Sugarcane Cutting (Piaui-Gastâo Vidigal-SP), 2011

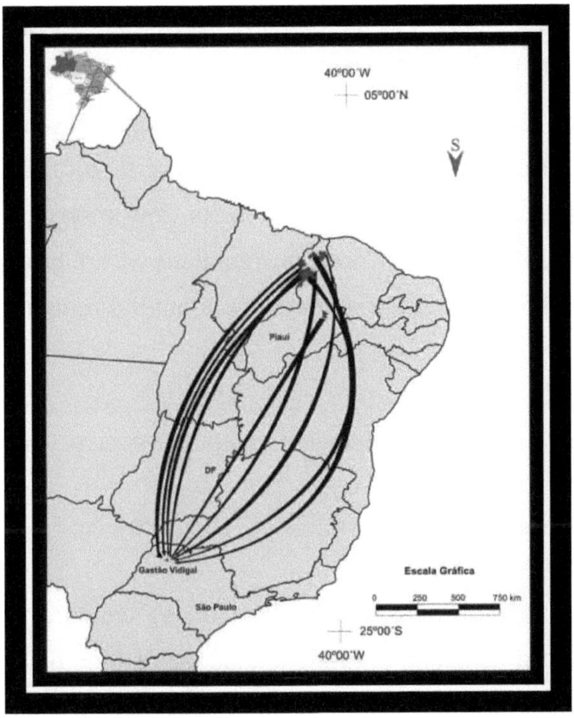

Source: Fieldwork/ Cardoso, 2011.

All right, we know where they come from, but one part of the question is missing: why do they come? What are they trying to achieve?

First of all, it must be made clear that migration, as Alves (2007) states, is a movement determined by *expulsion*, i.e. workers migrate when the conditions for reproduction in their places of origin are compromised. Therefore, migration on any scale is, as a rule, a socio-territorial process determined above all by economic and political aspects, and this statement takes shape when we study human mobility within the socio-metabolic system of capital.

As Singer (1979) points out in his analysis of migration, the factors that lead to migration are twofold: *factors of change*, which result, in a second moment, from the expansion of capitalist relations to portions of the territory not consolidated in this economy, generating expropriation of peasants, squatters, aggregates, partners and other forms of family farming; and *factors of stagnation*, according to him, resulting from the inability of producers in a subsistence economy to increase the productivity of the land.

According to Andrade (1981, p. 12), "migration is associated with an economic gap between points in the geographical space or the existence of a *new* area that is supposed to offer greater economic opportunities". In our view, migration is all of these things, and going further, through long conversations with migrant workers, we detected common threads in their statements about the reasons for their departure:

"I come here because the jobs are good, there aren't any there, and there are also people who come here for adventure, to build another house, a business to work there" (migrant from Piauí, 35 years old).

"I come here because I need money to survive, I come out of necessity, because the situation is difficult there" (migrant from Piauí, 20 years old).

As Oliveira (2009) states, the lack of means of survival drives workers to mobilize and imposes on them the condition of merchandise, since they have no alternative but to put their labour power up for sale on the capitalist market. Mobilization is the process that gives the worker mobility, and the mobility of the workforce is an indispensable element for capitalist exploitation. In this sense, as the migrant himself describes in his speech, his spatial displacement is motivated by *necessity, by the search to continue living.* It's not a mere exercise of his will, the spatial displacement in question is not a tourist flow, migration has nothing to do with freedom, or rather, it has to do with its loss. When asked what activities they did before migrating, the clues fit together and so we can understand why they came.

"Before, I used to work on my land, planting manioc, corn and other foods, so I didn't need much money, just to buy something at the market. Over time, the land became weak, I had no fertilizer, nothing, no help and no money, so I sold my land, and since then, I've come to work here cutting sugar cane" (migrant from Piauí, 29 years old).

I was a bricklayer's labourer, then the work became scarce, I'd go for a day, two, a week without a job, so I decided to come to São Paulo to try and make a life for myself" (21-year-old migrant from Piauí).

In these terms, what we see from the migrant workers' testimonies is that a large part of them had their livelihoods based on the land, on small properties, and that when they were uprooted, dispossessed of their territories, the only path left to them was the one stipulated by the system of capital's social metabolism: to move around and sell their labor power as a commodity. However, in today's world of work, marked by neoliberalism and flexible production, instability and unemployment are not just cyclical, but have become structural and so, without a job in their region of origin, they see cutting sugar cane as a way out of this problem.

In other words, in general terms, these people migrate either because they don't have access to land, or because they can't find a job in their place of origin and have a family to look after, and in such a situation the worker ends up seeing migration as the only alternative for survival, so this migration is therefore forced, driven by the need to survive.[8]

But the situation is not only difficult for those who leave, but also for those who stay. This is the case with women and children, because with the departure of their husbands or fathers, they are the ones who experience the drama of scarcity and hunger most closely, especially when the migrant is unable to send a little money home. The expression "widows of living husbands" demonstrates the strong rupture caused by migration within families[9] . Of the total number of people interviewed, approximately 95% come to cut sugar cane without their families, leaving their wives and children in the state of Piaui. In this respect, their testimonies are full of sadness and nostalgia for their loved ones and their land.

"My wife and children stayed there, I couldn't bring them, the ticket is expensive and you never know what you're going to find here, I miss them, but I have to stay here and earn a bit of money for them and me to survive" (26-year-old migrant from Piauí).

---

8 For more information, see GUANAIS, 2008, and ALVES, 2007.

9 Cf. Migrantes Temporaries, 1991; Debate on Migration and Citizenship, by the Pastoral Migratoria de
Guariba, 1997.

"The family stayed, I couldn't bring them, I didn't even know what it was going to be like here, every month I send a little money to the wife and children, I miss them, I won't be able to see them until the end of the year, I miss my land, the people from there" (28-year-old migrant from Piauí).

Among the migrants, there is a strong sense of belonging, emotional ties to the "place" of origin and the desire to return to it; the place of origin is seen as a good place to live, but the social conditions don't allow it[10] . The migrant workers evoke in their speeches the desire to return, the feeling of longing, but for them there is no going back before the end of the harvest, there is not the slightest alternative of returning without money, the pain, the suffering are experienced and silenced daily, the ordeal has to be endured, in their minds going back with nothing is like attesting that they are weak.

The arrival of migrants in Gastâo Vidigal-SP basically takes place in mid-March of each year, intensifying in April, and they remain until the end of the harvest at the end of November. As part of the dynamics of this seasonal work process, at the end of November and the beginning of December, an important part of the migrant workforce returns to their regions of origin to reunite with their families. We also found out during the course of our research, in conversations with the migrants, that during these periods there are frequent chartered buses taking them back to their cities of origin. Very often, during the journey, these buses are ransacked by robbers who already know the migratory flow route and take what little they have managed to save during the harvest with great sacrifice.

We've already told you where they come from, where they go and why they go. Now we need to know how the migrants come to cut sugar cane in Gastâo Vidigal-SP, how they are recruited and brought in.

If I'm not mistaken, the migrant workers are recruited and brought to the municipality by "cats". The system works as follows: the "cat" resident in the municipality personally calls or commissions a "sub-cat", most often a migrant

10 Cf. Pereira, 2008.

himself who is respected by the other migrants, to bring in the number of workers requested by the first.

In a conversation with a "migrant sub-cat", we found details of this process of precarious recruitment of the workforce for sugarcane cutting:

"I'll call the boys on March 5th and they'll be here on March 8th. In my class there will only be 'good peasants', of R$1000.00; R$1200.00". (Piauían migrant sub-cat, 35 years old).

Still on the subject of recruitment, it is at the moment of departure for the sugarcane fields that the saga of the overexploitation of these migrants begins, because all they have to do to make the journey is get into debt, as the migrants themselves say:

"You have to have the money when you leave, so some people sell what they have to come here, they sell refrigerators, televisions, radios, anything that can turn into 'money' to cover the cost of the ticket." (migrant from Piauí, 26 years old).

It sounds almost ironic, if it weren't for the seriousness of this socio-territorial problem, since migrants are not only overexploited in the work process, but also outside of it, as soon as they leave, they have to make their social condition even more precarious, giving up their possessions acquired with the sweat of their brow, in order to buy the ticket that could lead to their death.

Therefore, the only certainty that all the migrants interviewed have, and which is a hallmark of the migrant workers' discourse, is that if social conditions were different and allowed them to live in their regions of origin, they would never have joined the living wheel of the migration process, because nobody migrates because they want to, nobody leaves their family, wife, children, friends, just because they want to, because of some act of "freedom", in other words, you don't choose to be a migrant, but rather, in the face of the destructive logic of capital, you are forced to be one.

"If things were different, if we could at least live and take care of our children in Piaui, I would never have come here, nobody leaves their home unless they really need it" (migrant from Piaui, 23 years old).

The vast majority of the population, especially the sectors condemned to social exclusion, leave their land and their people not out of free will, but for reasons of life and death. Their very survival is at stake. Hence our insistence that the right to come and go is matched by the right to "stay". How many migrants who travel the roads today, if they had the choice, would decide to remain in the soil where they buried their dead! Migrating must be a free decision and not one forced by survival (GONÇALVES, 2001).

Therefore, to the more unsuspecting, migration, the coming of northeasterners to cut sugar cane, is seen by some scholars as an act of realizing their freedom. In this work, I defend the opposite idea, that the act of migrating does not mean and does not imply the de facto freedom of the worker, "freedom" in these terms is just a *dream;* the de facto reality is *forced migration for capital.*

But there is a question in the air: what is the meaning of the vaunted *freedom* expressed in the relationship between capital and labor? To what extent are migrant workers really free? To what extent does the freedom postulated by the capitalist order actually refer to the freedom of the worker? The question that haunts and motivates us: would the migrant have the freedom not to migrate, to refuse the exploitative contract with the agro-industries?

What is important to emphasize beforehand is the way in which bourgeois society is concealed by the ideology of freedom and equality. The French Revolution was the political and social milestone through which these principles were disseminated. In this way, a set of abstractions and inverted ideas about the new social order that was about to be established was gestated at the level of representations. According to Silva (2008), these two principles had in fact been built up since the 18th century by the Enlightenment, principles that envisioned the construction of another society, different from the existing one. The first step was to guarantee free men to work, without the marks of slavery and servitude; the second was to guarantee equality to mask the profound inequalities. Equality, as a principle, was based on the equalization of commodities as exchange values. As mentioned above, these principles established rights, which were not actually given, but were only an ideal to be pursued.

In this way, the freedom thought of by the capitalist bias would be achieved by the much-vaunted economic "development", the expansion of the market economy and wage labor, as Sen (2000, p. 21) states, "it turns out that the rejection of the freedom to participate in the labor market is one of the ways of maintaining the subjection and captivity of labor". In this statement, it is clear what kind of freedom the system of capital proposes, that is, the *pseudo-freedom* postulated by the social order of capital only guarantees freedom in the sense of building a labor market, it only guarantees the buying and selling of labor power, a fundamental mechanism for the expanded reproduction of surplus value. As such, capitalist freedom boils down to the realization of surplus value. The social being is said to be "free" as long as he is only a commodity for the sale of his labor power.

For Prado Júnior (1986), capitalist freedom consists of the establishment and acceptance of agreements with other individuals, and this is where freedom goes in the bourgeois regime. However, this equality of individuals in the freedom to agree with each other is, nevertheless, a legal equality, that is, a freedom of law and not of fact. Because the reality is that individuals are very unequal, and particularly in what most contributes to setting the limits and scope of their action, namely their position within the social structure, it follows that the freedom of each will vary greatly depending on the real inequality created by capitalism.

> In this sense, bourgeois freedom is nothing more than an illusion, since it grants the individual a power that the contingencies of collective life immediately take away. Or they take it away, in most cases, from the majority of individuals. The individual is "free" to choose and determine his action, but when he tries to carry out and make his choice effective, he finds that the contingencies of social life, determined by the "free" choice of other individuals who are better off than he is, gradually narrow down his initial "unlimited" freedom and restrict his action until it is reduced to a minimal sphere to which he will find himself ineluctably restricted and condemned (PRADO JÛNIOR, 1986, p. 59).

In the sense in which we understand freedom, in the face of alternatives (in the plural), of opportunities for the social being to realize himself, to give vent to his potential and to set his aspirations according to them, achieving them, the means to choose the desired way for workers to carry out their life projects, or rather, freedom as the suppression of the alienation enclosed by capital, I defend the idea that migration to the cutting of

This is not an act of freedom, on the contrary. Because we agree with Basbaum (1967, p.113) when he states that: "A

Freedom is the suppression of alienation; the suppression of alienation is the suppression of the system; the suppression of the system is the suppression of private property, the source of all alienation."

We are not defending the idea that we live in a slave or feudal society; the capitalist social formation did experience a process of wage-earning, in which the human force that works was stripped of its means of production. However, we are pointing out the confusingly perceived ideological fog that covers the meaning of being "free" in capitalist society. First of all, without any beating around the bush, we can state that *freedom* does not actually exist in capitalist society, it is just a fallacious notion of bourgeois legal ideology. Within the social metabolism of capital, freedom has no place; social relations anchored in alienation predominate.

Therefore, instead of being a free decision, the right to migrate becomes a compulsory obligation for survival: "coming and going", far from being an act of freedom, ends up revealing the desperation of those who find themselves pressured by the need to survive. Thinking about the situation of the real heroes of this immense "sea of cane", the rural workers, especially the migrants, we can venture the hypothesis that they find themselves in conditions of "non-freedom", that is, they have no options or choices. Thus, the *pseudo-freedom* offered by capital completely disregards the development of human omnilaterality, it is a *"freedom for capital"*. In this contradictory movement of this civilization of barbarism, migration to cut sugar cane becomes *"migration for capital"*. In other words, we are not against the right to come and go, but in favor of the right of men and women to be able to experience space, to be able to freely choose the spatial arrangement where they want to settle, and in this sense, it is also essential to guarantee the *"right to stay"* in dignified conditions in their places of origin.

CHAPTER III . THE DEGRADATION OF SUGARCANE WORKERS: MIGRANTS' WORKING AND LIVING CONDITIONS

In order to better understand the complex universe of job insecurity and exploitation to which sugarcane cutters are subjected, we need to know and understand the work process in sugarcane cutting from end to end, because only then will we be able to grasp the clashes that are being waged in the midst of the capital vs. labor relationship. It is only in this way that we will be able to understand why cane-cutting workers die, fall ill and suffer?

Alves (2007), Silva (2007) and Thomaz Junior (2007) make important points about labor relations in sugarcane plantations, each with their own specific directions, focusing attention on the intensification of exploitation mechanisms and the worsening of workers' quality of life.

Starting with the work process of cutting sugar cane, we can see that it consists of cutting all the cane in a rectangle - called eito by the workers and contained in the plot - which only has its width defined in advance, since the length depends on the cutting capacity of each worker and therefore will only be observed at the end of the working day by the scorer (inspector), Therefore, it will only be observed at the end of the working day, when it will be measured by the scorer (inspector), it can be seen that the intensity of the work is predetermined according to the potential of the worker's strength and dexterity, but which in turn is established by the company.

The prevailing width of the eito in the state of São Paulo is five streets[11] , equivalent to 6 meters between the end rows, with a spacing between one row and another of 1.5 m. Cutting begins with the central row of the eito (3ª row), where the sugar cane cut from the entire eito will be deposited[12] . (Figure 1).

---

11 However, this was not always the case, as this configuration was achieved by the workers during the Guariba Strike in 1984. For more details, see: Revista Proposta, n°42, 1989; Thomaz Junior, 2002.

12 It is important to note that in each meter of a cane row in São Paulo, in first cut plantations, there are between 9 and 18 cane stalks.

Figure 1- Sugarcane cutting scheme

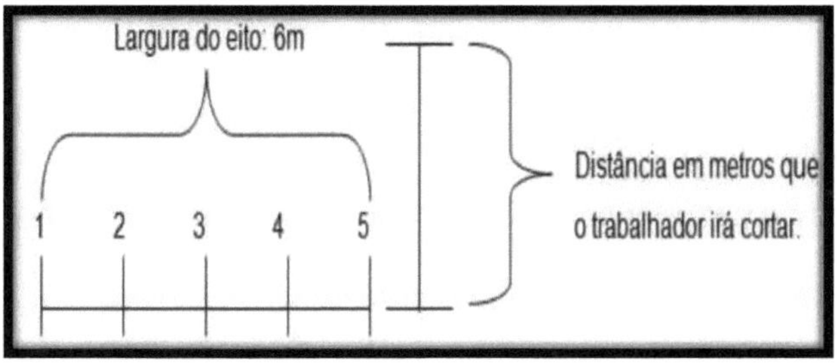

Source: Cardoso, 2011.

Once all the canes in the bundle have been cut at the base, the worker cuts the tip, i.e. the top of the cane, where the leaves are. In some sugar cane companies, the workers have to cut the tip in the air, in others it is allowed to be cut on the ground, in the central row; in some the cane only has to be cut in the central row, but in others it is obligatory for the workers to leave the cane in piles, one meter apart, the so-called "flags". In this sense, we can see from the testimonies of the migrant cane cutters that they don't make the decisions about what to do, where to work and how to carry out the activity; the technical requirements for cutting, which are demanded by the agro-industries, are appropriated by the workers and integrated into their way of working.

to operate.

"It's like this, we don't decide which way we're going to cut the cane, it has to be the way the company asks, very close to the ground, don't leave any high stumps, where I'm standing I have to remove the tip, it has to be a very small tip and still high up, so as not to dirty the cane too much. To finish, we have to make a pile every 4 meters, which means more time and work for us". (migrant from Piauí, 25 years old).

We therefore agree with Iamamoto (2001) that cane cutters have no control over the *content of* their work, since when they start working they are already part of a global production mechanism that is external to them.

Thus, work in sugarcane cutting is not limited to the activity of cutting cane, i.e. removing it from the ground using an instrument, the machete (or pruner). The work of cutting cane involves a number of other activities: a) cleaning the cane and removing its tips; b) transporting the cane to the center line of the eito and; c) arranging the cane, deposited on the third line, for mechanical loading. Therefore, for the workers, carrying out these additional tasks doesn't correlate to additional earnings, but it does mean spending additional energy and reducing the time that could be used for cutting cane and earning more income.

The experience of working in the sugarcane fields has been dominated by a double characteristic: long working hours and the maximum intensification of work, stimulated by payment based on production, i.e. the amount of cane cut each day. This double mechanism is at the root of the *suffering of the work.* The migrant workers say that the working day is prolonged by the journey to the unpaid cane field, and the risks of transportation to the site.

"Ah, I get up at 4:30 in the morning, make some food, put some cold water in the bottle, pack my backpack and go to the bus stop with the others to wait for the bus. At 5.30 a.m., the bus passes by, I get on, and it takes more than an hour to get to the sugarcane fields. By 4:30 p.m. we leave, take another hour's drive and arrive home by 5:30 p.m., already very late. And those two hours we spend on the journey, we don't earn anything". (migrant from Piauí, 28 years old).

"The buses that take us are very ugly, they're old and there's no safety at all, the seats are hard and some of them are broken, even a while ago the speedometer was broken, it didn't register anymore." (migrant from Piauí, 23 years old).

According to the migrants' accounts, the working day, including the time spent moving to the cane fields, preparing food and waiting for transportation, averages 11 hours. This means that the worker starts to see most of his time as time that doesn't belong to him, dedicated to working for someone else and, what's worse, a large part of this time doesn't come back to him in the form of wages, as the testimony says the hours spent traveling aren't being paid. We can also interpret from the reports that they make the most of their working time by dedicating just a few moments to their meals, taken during transportation or when they arrive at the cane fields. More than just a cultural habit, the routine of having lunch early expresses a wisdom, drawn from experience, in which the meal provides them with the energy they need to face this highly strenuous work. If that weren't enough, the workers still live with insecurity on their daily commute to work. The fleet of buses that transport these workers is in poor condition, and to disguise it, the drivers try to make up their vehicles by painting their façades; on the outside the sensation is pleasant, on the inside the reality is precarious.

Thus, the workers with the highest productivity are not necessarily those with the greatest muscle mass, but rather those with the greatest physical endurance to withstand the intensity of this repetitive and exhausting activity, carried out under the sun, therefore, high temperatures - even if it is concentrated in the fall and spring seasons - in the presence of soot, dust and for a period that varies between 8 and 11 hours of work.

In these terms, the sugarcane agribusinesses installed in municipalities neighboring Gastâo Vidigal-SP, given that the municipality under analysis does not have any units installed in its territory, have been making use of the search for migrant labor. For capital, the migration of workers living in precarious conditions is a primary source of its high profits and of political power and influence, through which it can easily raise public funds from the state. In addition, we can say that migrant workers are preferred by the companies not because there is a shortage of local labor, but above all because they are more physically resistant and more "submissive" to the conditions imposed. Mill representatives have already made it clear that they prefer to hire workers from states (and regions of states) in the Northeast, who are less politicized, i.e. more susceptible to the dictates of capital and therefore less threatening from the point of view of claiming rights, than local workers.

The fact is that most companies are making use of migrant labor, on the grounds that there is a lack of local labor on the expansion fronts. One of the main problems faced by the municipality of Gastâo Vidigal-SP is the unemployment of its population. It should be noted that in addition to these factors, many companies prefer to hire migrant workers because they believe them to be more productive and less demanding than local workers. They prefer migrants because they don't claim anything and work quietly. In this way, it is worth emphasizing that Northeastern workers are preferred by mills in the Center-South, not because there is a shortage of local workers, but above all because they are more physically resistant.

Migrants are extremely productive workers compared to other workers. Due to historical issues - cultural, economic, social or even religious - and a trajectory of work experiences or labor sociability marked by high levels of precariousness, they hardly refuse "hard work".

"You have to face up to the work, there's no other way out, no matter how difficult it is, it's the only thing you have to do, or you have to put up with everything, or fail and leave your family with no money, going hungry, needy" (migrant from Piauí, 27 years old).

We can't ignore the fact that within the social and sexual division of labor established by capitalist society and reinforced by bourgeois/Christian ideology (patriarchal structure), men have a well-defined role as the father-provider and head of the family. In the workers' daily lives, the fear of failing to fulfill this role and the punishments coming from the social environment are decisive objectivities for the construction of forms of subjectivity or even social self-representation, which is why there is a certain rigidity *in self-discipline* in order to endure "more work" under worse conditions. Because they are migrants from distant regions and because they are the result of a process of expulsion in their regions of origin, they are not expected to show any disagreement with the demands of the sugarcane agro-industry.

Another characteristic and requirement of cane-cutting work is the youthful nature of the workforce required, as we saw during the interviews with the migrant workers, the age range is between 18 and 35, with an average age of around 26. It should be noted that among the migrant cane-cutting workers, no women were found during the research, and the number of local women in cane-cutting is falling. Today, the majority of companies are restricting the hiring of women and some of them are directing female labor to activities other than cutting cane. In this sense, Silva (2006) points out that:

[...] they are thrown out of sugarcane cutting under the pretext that they don't meet the required productivity levels. Those who do get work are being relegated to the worst tasks, such as "bituqueiras" (which consist of collecting the "bitucas", i.e. the pieces of cane that are not picked up by the machines after cutting), or as poison applicators to control cane pests (activities known as "descarte"), or in the nurseries where the buds are prepared for planting cane, which also require the handling of pesticides [....] (SILVA, 2006, p. 64).

However, in order to fully understand the work process to which the workers are subjected, it is necessary to understand how they are paid, i.e. why the pace of work is accelerated and intensified?

This rhythm is accelerated and intensified as a result of an element external to the production process, which is *payment for production.* This form of payment acts on the worker's psyche and pushes him to increase work productivity, in other words, to intensify it. Thus, Iamamoto (2003) points out that this form of payment encourages competition between workers and reinforces individualism; because "on the eito it's every man for himself and God for all", although the harvest is only carried out with the work of many workers, the activity of cutting is encouraged to be experienced individually and in solitude. And yet, under this system and given the demobilization of the workers and the inefficiency of the unions, the workers generally only know how much they have earned on a given day when they receive their pay *slip* for the month, fortnight or week. Not to mention that they only know how many meters of cane they cut in a day, but they don't know the value that will be assigned to the meter of cane on that day, because this value will only be determined at the mills through a conversion.

Under these conditions, the sugarcane agro-industrial units weigh the cane cut by the workers and assign the value of the linear meter through the relationship between the weight of the cane, the value of the cane and the meters that have been cut, all, of course, without the participation of the workers, thus enabling the recurring practices of fraud and theft against the worker[13] . During the

13 Due to the mobilizations of the workers and the Rural Employees Union (SER) of Cosmópolis has a closed court system, under which the workers are able to extend their attention

interviews with the migrant workers, they all said that they prefer to be paid by production in order to get a higher wage, but they also understand that they could earn much more if it weren't for the companies' scams to cheat them out of their pay.

"I think there's theft at weighing time, because at that moment there's no one accompanying the workers to check on them" (migrant from Piauí, 22 years old).

Payment for production in sugar cane cutting is something terrible and controversial. In the interviews with the migrants, practically all of them said that they prefer payment by production in order to achieve a higher salary, but they also understand that they could earn much more if they weren't cheated by the companies when they were *paid.* But the wage form makes any division between paid and unpaid work invisible, appearing as if all the work created is paid for, which leads workers to believe that the wage pays for all the work done, depending exclusively on their commitment. The fetishism of the wage form is reinforced by payment for production. In other words, payment for production hierarchizes and divides workers, so that in order to get workers to compete with their peers and increase production, it is necessary to instill in workers the values of bourgeois ideology, such as individualism, meritocracy, belief in justice, loyalty, honesty and self-esteem.

Regarding payment for production, Adam Smith in the 18th century and Karl Marx in the 19th century already denounced this form of payment as one of the most inhumane and perverse; it is true that they denounced this work, analyzing this form in situations where the worker at the end of the day was fully aware of the amount he had earned. At the moment, while the agribusiness discourse is growing stronger and the process of productive restructuring is intensifying in the country, the precariousness of work in the sugarcane fields is reaching impressive levels, not forgetting that what Thomaz Junior (2002) found in the 1990s is being consolidated in this context, i.e.

---

to all the work operations involved in cutting, loading and weighing the scales at the Ester mill.

workers are being forced to follow the productivity standards imposed by the harvesting machine.

The workers' complaints and various studies from different areas[14] are revealing to society as a whole the existence and imposition by sugarcane capital of extremely high manual cutting targets to be met by the workers, and today there is a system of "pruning" in the mills. Pruning is as follows: if a worker doesn't manage to cut a minimum of 9.10 tons a day, the mill cuts him off, i.e. prunes him before he has completed two months of work.

"Yes, there is a target that we have to reach, in this case it's around 10, 11 tons, if we don't reach it, in the next harvest the mill won't accept you anymore. I think I get an average of 15 tons, but there are really good peasants who get up to 20 tons a day". (migrant from Piauí, 26 years old).

It should also be noted that in many cases, failure to meet these targets is turned against the worker himself, becoming an obstacle to his hiring for the following harvest. Worker productivity is a powerful instrument for controlling the workforce in the hands of sugarcane agro-industrial capital, both at the time of hiring (when the harvest begins) and during the harvest (THOMAZ JUNIOR, 2009).

Thus, the imposition of minimum productivity can be seen as a kind of moral coercion, consisting of one of the possible forms of the current reinvention of the exploitation of labor by agribusiness, what Thomaz Junior (2002) called the *standard of cutting in the cut*. The same can be seen with labor productivity: while in the 1970s labor productivity was, on average, 3 tons, in the 1980s the average cut by workers rose to 6 tons, in the 1990s this average rose even more, to 9 to 10 tons/man/day and currently this average is hovering around 12 tons. Therefore, average productivity quadrupled between the 1960s and the beginning of the 21st century.

In the midst of all this domination and control, the aim of capital is for workers to incorporate its logic and legitimize it, making the company's objectives their

14 Cf. THOMAZ JR, 2007; ALVES, 2008; GUANAIS, 2008.

own. Thus, local sugarcane capital uses certain strategies to this end, such as illegal practices like the "prize of the week" or "golden pruning", which consists of the mills rewarding the most productive worker with a "gift" (from cell phones to motorcycles).

"I won a cell phone last month for being one of the best cutters in the class, this is good for us because it encourages us to work, the boys are already taking it harder, they want to beat me". (migrant from Piauí, 21 years old).

In this respect, Thomaz Junior (2002) highlights the strategy used by capital to make it possible to intensify the pace of work: the announcement, at the start of the harvest, of "various prizes" (cars, bicycles, radios, fridges and other household appliances), This process, as the author calls it, is called "*Death Bingo*", since the worker could be the next victim of this unfair "lottery", since he will go to any lengths to win the "prize". In other words, from the migrant worker's statement above, we can say that sugarcane capital has managed to capture and control the subjectivity of workers, inserting the logic of capital into their way of thinking, legitimizing the over-exploitation of labor to which they are subjected. This strategy guarantees capital an air of "kindness", as if it were supporting and honoring labor.

However, in addition to the increase in labor productivity, there was also a brutal reduction in the wages paid to the cutters. According to surveys by Alves (2007), in the 1980s, following the series of strikes that began in Guariba in 1984, the workers won a wage floor for the category of two and a half minimum wages. The wage floor of two and a half minimum wages would be equivalent today to R$1,362.50 (considering the minimum wage of November 2011). However, according to data from Alves (2008), the salary floor for the category in São Paulo is around R$ 460.00, which means that the salaries practiced today are 2.96 times lower than those practiced in the 1980s. In Gastâo Vidigal-SP, the average wage for migrant workers is around R$ 900 reais, as we saw during the interviews, which is below the floor agreed in 1984. Therefore, it is absolutely clear that between the 1980s and the 2000s, there was a strong wage squeeze in the category of cane cutters and, on the other hand, a strong increase in labor productivity in the cane fields.

Therefore, for capital to continue its reproductive cycle in this world system in which competitiveness is the guarantee for the survival of companies, it is essential that it uses modernized forms of production and, at the same time, resorts to precarious and intensified forms of exploitation of labour. As Antunes points out:

> It's one thing to have the imperative need to reduce the variable size of capital and the consequent need to expand its constant part. It is quite another to imagine that by eliminating living labor, capital can continue to reproduce itself (2001, p 120).

We can say that, although the frequent discourse of businessmen in the sector, as well as representatives of public authorities, is that the expansion of sugarcane activity will bring benefits to the region, such as, for example, job creation, land appreciation and a boost to the local/regional economy, consequently leading to greater regional economic growth, it is necessary to pay attention to the circumstances in which these jobs will be generated, what the working conditions will be for those who are already employed, as well as for those who will be included in this process.

In the current context of the valorization of capital, which, when realized, reveals the face of barbarism, the overexploitation of labor is expressed with greater intensity in the case of migrant workers. In this scenario, the Public Ministry of Labor has detected countless cases of irregularities in terms of employment contracts and, especially, the material conditions for workers' survival, which are essential for the revitalization of their workforce and a right guaranteed by law.

In Gastâo Vidigal-SP, the situation is no different. In 2005, inspectors from the Ministry of Labor and Employment (MTE) found more than 100 migrant rural workers from Piauí, living in subhuman conditions in the municipality's lodgings. In 2007, a group of rural workers recruited in Piaui denounced working and living conditions incompatible with NR-31, on two sugar cane farms owned by former Agriculture Minister Antonio Cabrera Mano Filho. The complaint was made to the Ministry of Labor and the Public Ministry of Labor. The workers said they had been brought to work on the Sâo José farm, in Gastâo Vidigal-SP, where 22 of them were living in precarious conditions in three abandoned houses. They claimed that everyone lacked work equipment and that there was sometimes a shortage of water, even for individual consumption.

With regard to the living conditions of migrant workers in the municipality of Gastâo Vidigal-SP, the mobility of labour forced by the loss of the means of survival and influenced by the dynamics of capital in its process of valorization leads to deterritorialization and, consequently, to a new territorialization. However, as Oliveira (2009) points out, the conditions in which this new territorialization takes place are as precarious or more precarious than those which resulted in deterritorialization, given that under productive restructuring and the globalization of capital, work is increasingly precarious, fragmented and multifaceted, leading to precarious *reterritorialization.*

With regard to housing conditions in the places of reproduction, we were able to visit the huts and shacks that serve as shelters for the migrant workers during our research. In general terms, in the municipality, these are rented in the name of the "cat" responsible for the group, The latter is also responsible for providing and arranging mattresses, beds and a fridge, almost always all of which is used. On the other hand, the migrants commit themselves to buying food and various products at the "gato's" own store, "gato's son" or any other relatives at the "gato's" pleasure. Thus, the relationship of domination and alienation of these workers is evident on a frightening scale, as we can see in what they say:

"I've been buying at the market around the corner since I arrived, I've never gone anywhere else to buy, I don't know if it's cheaper elsewhere, we buy there because it was the market that the cat pointed out when we arrived." (migrants from Piauí, 28 years old).

This means that this strategy is adopted when irregularities related to the agency, hiring and maintenance of labor in precarious conditions are discovered, so that the sugarcane agro-industrial companies are exempt from any blame. When this happens, it is also discovered that the processing units are the ones who renovate the lodgings, put the workers up in boarding houses or rented houses and pay for everything. It is easy to hold third parties responsible, especially when it comes to the "cat", because he is always hiding from inspection, given that his actions are considered a crime, according to article 207 of the Brazilian Penal Code, which states: "Enticing workers, in order to take them from one place to another in the national territory. Penalty - imprisonment from 1 (one) to 3 (three) years, and a fine". (OLIVEIRA, 2009).

Migrant workers who have been expropriated from their land or deprived of basic subsistence conditions with their families are transformed into subjects whose human condition is denied to them when they migrate. There are lodgings that don't have adequate facilities for workers to sleep, store their belongings, eat or have any kind of leisure time when they're not working. As for the migrants' living conditions, the houses are mostly small, with 2 or 3 rooms, where up to 6 people are crammed into each room. Despite all the exploitation carried out by capital, i.e. sugarcane agribusiness, in the working environment, in the very spaces where life is reproduced and regenerated, the exploitation of capital over labor continues to be a constant in the lives of migrants, condemning them to such a subhuman life even outside the working spaces. (Photo 1).

Photo 1- Migrant Housing Conditions in Gastâo Vidigal-SP

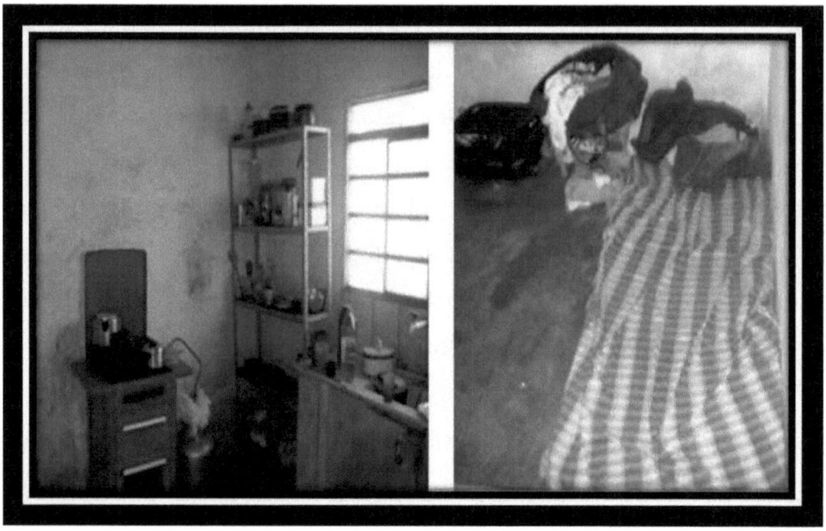

Source: Fieldwork/Cardoso 2011.

Most of the houses we visited, as well as physically looking unkempt and/or dilapidated, had irregularities such as workers without beds and sleeping on torn mattresses on the floor, food (from the food basket) piled up on the floor along with the mattresses, clothes, lack of hygiene in the environment, dirty kitchens and walls; clothes strewn everywhere, stoves and gas cylinders in a precarious state of use next to the bed or sleeping mattress, posing a risk to the workers; dirty pots and pans of food strewn all over the floor.

Thus, as Silva and Menezes (2000) point out, housing is an extension of what we see on the sugar cane fields. Crammed into dark, poorly ventilated, dirty rooms, these men feel the weight of being cheap commodities in a world where rights only exist on paper (Photo 2).

Photo 2- Precarious migrant living conditions in Gastâo Vidigal-SP

Source: Fieldwork/ Cardoso, 2011.

These facts help to corroborate the thesis that capital feeds on modern forms of production and management of the production process and work, but maintains precarious forms of work, often masked by the discourse of adopting economically viable, environmentally correct and socially just practices.

What's more, in the course of our research we found that because they are far from their families and isolated from the social practices developed by the population as a whole in the cities where they are staying, these workers have no leisure options on their days off and, above all, at weekends, leaving them with two options: those who like it go to the bar to drink alcohol (or take it home) and the others just watch television. This was stated by several of the migrant workers interviewed. In some of these lodgings, we spotted several empty bottles that show the consumption of alcoholic beverages. Although the workers are not shy about confirming that several of their colleagues drink alcohol,

They were keen to point out that this usually happens at weekends.

It is important, however, to emphasize that when we refer to the use of alcohol by these workers, we do not intend to denigrate their image, but to point out that this is yet another consequence of their social condition, of the deterritorialization that these workers suffer and of the precarious living situation to which they migrate.

> Thus, far from their families, isolated from the local community, under pressure at work and under the control of the contracting companies and with no leisure options, these workers are vulnerable and find alcoholic drinks as a way of "venting" their dissatisfaction (OLIVEIRA, 2009, p. 439).

We can see that, in general, precariousness prevails, both in working and living conditions, with capital subjugating and controlling both working spaces and the regeneration and reproduction of the workforce. As we had the opportunity to see in *the field*, there are many cases in which migrant workers live practically crammed into small spaces, in extremely poor hygiene conditions. This brings us to another crucial point for analyzing the category of work: that a life full of meaning outside of work presupposes a life full of meaning inside work. It is not possible to reconcile paid and alienated work with truly free time, or even with real freedom. A life devoid of meaning at work is incompatible with a life full of meaning outside of work. Therefore, by analyzing the working and living conditions of migrant sugarcane workers, we can affirm that, under the logic of capital, the worker is not satisfied at work, but degraded, not recognized, but denied.

> The worker therefore only feels close to him/herself (when) outside of work outside of him/herself (when) at work. They are at home when they are not working and, when they are working, they are not at home. His work is therefore not voluntary, but forced, compulsory work. Work is therefore not the satisfaction of a need, but only a means of satisfying needs outside of it (MARX, 2004 *apud* ANTUNES, 2005, p. 83).

After all, in the face of this process of alienation of labor promoted by capital, the migrant worker only feels free in situations typical of animality (eating, drinking, procreating); in his human functions, he feels like an animal, excluded from society.

CHAPTER IV . MIGRATION AND LABOR CLASS (DIS)BELONGING

In order to develop this chapter, we first need to be minimally aware of the debate between those who preach the end of work and others like us who defend the maintenance of the centrality of work, with the metamorphosis and complexification of the working class today. To start this conversation, we can point out that Kurz is perhaps one of the most enthusiastic proponents of the thesis of the end of work, in which he argues that there are no subjects capable of leading the emancipation of society. In other words, the thesis of the crisis of the society of work, by denying the political centrality of work, does not envisage any emancipatory possibility, other than that led by the organized actions of the non-working class, along the lines of Gorz's original formulation, or the social movements that are inscribed beyond the framework of work. In other words, if something has to happen to shake up and break with the metabolism of capital, it won't be in the trenches of work.

We would like to defend the opposite idea, that of the primacy of the centrality of work on the threshold of the 21st century, which obviously doesn't mean that its morphology hasn't changed profoundly in contemporary society. It is, of course, a question of understanding these changes, but we must not lose sight, among other things, of the fact that abandoning the centrality of work in explaining contemporary society also implies abandoning the labor theory of value.

As Thomaz Junior (2009) states:

> We can say that betting on the infertility of the non-centrality of work, even though two thirds of humanity live the scourge of precariousness, marginalization/exclusion and all forms of subordination/domination/expropriation/subjugation, of unemployment, is the same as not being able to go beyond what is visible. and all forms of subordination/domination/expropriation/subjection, unemployment, is the same as not being able to go beyond what is visible, or betting on the inability to grasp the objective and subjective contradictions of the social structure in force. (p.112).

Following Thomaz Junior (2009), in some circumstances, the debate on the very centrality of work assumes the reductive identification of work with the worker (salaried) and, consequently, the work-employment confusion, as Gorz's entire formulation is rooted in, when, in fact, it is not work that has ended or is ending, but *employment that is dying*. Or rather, it's capital that is using its workforce in a different way, because if in the past "the form of direct salarization prevailed, today it is possible to observe that, through outsourcing, self-employment and entrepreneurship are being encouraged".

In the same vein as the de-centering of the category of work, we have the idea of Habermas or his theory of communicative action, just pointing out that the duality between the system (instrumental reason) and the world of life (intersubjective space, the sphere of freedom) is the basis of the author's understanding that work is being progressively displaced by science and technology, undermining and emptying the Marxian argument of labor-value, since science becomes the main productive force.

Following the lead of Antunes (2005) and Thomaz Junior (2009), we believe it is essential to consider work in all its dimensions, particularly when dealing with the crisis of the labor society. In other words, this subject requires us to explain which dimension we are talking about: whether it is a crisis of abstract work or of work in its concrete dimension, as a structuring element in the social exchange between men and nature. The crisis of labor is therefore linked to the crisis of abstract labor, or the form of being of labor under the reign of commodities, which takes on a strange character under capitalism.

In view of this, we agree with Thomaz Junior's (2009) view that it would be a mistake to propose the end of work or the loss of its centrality as long as capitalist society persists.

There is thus a dialectic of work (negativity/positivity) which, if ignored or neglected, radically resizes the debate around its centrality and, consequently, its emancipatory power. Of course, it would be impossible to imagine the elimination of work or, to a certain extent, of the working class, as long as the constitutive elements of the capitalist societal structure remain in force. Through the Marxian perspective, we put this question back in the following terms: the overcoming of abstract work and, in its place, the validity of concrete work, linked to the production of use values or goods for the satisfaction of needs, the product of which would enable free time, the materialization of a meaningful life and human emancipation (p. 118).

The scale of the controversies that this issue contains in the Marxist sphere is beyond our objectives in this text, but it is important to point out the arguments and counter-arguments, without our intention escaping the recognition of the explanatory limitation - to which we also subscribe - the desire to discuss a possible *broadening* of the concept of work and the working class for the reality of the 21st century and the emancipatory and revolutionary desire.

What is the current shape of the working class, at least in its most generic form?

Without a shadow of a doubt, today's working class has a new morphology, its formation is even more fragmented, more heterogeneous and more complex than that which predominated in the heyday of Taylorism and Fordism.

In this respect, a first point of disagreement can be seen in the position defended by Lessa (2007), who, in several passages of the text "Trabalho e proletariado no capitalismo contemporâneo" (Work and the proletariat in contemporary capitalism), expresses his fidelity to Marx's original formulations, particularly Book I and, in this case, argues that "the origin of all the wealth that circulates in society is labor, more precisely, proletarian labor [....], is the only one that produces a new *quantum of* the material content of social wealth, which will be added to the amount of global social capital, therefore, the only class that performs the social function of converting nature into means of production and subsistence". In short, the author is basing his understanding on the fact that the working class is made up only of the proletariat.

Compliance with the rigidity of Marxian formulations raises the defense of objective class contradictions between the proletariat and other wage earners. Lessa (2007) goes further when he argues that the fact that non-proletarian wage-earners live off the wealth that the bourgeoisie expropriates from the proletarians encourages them to maintain the defense and maintenance of capitalism, because "they act predominantly as an auxiliary force in the reproduction of capital", particularly with their attachment to private property, a historically recorded fact whenever this has been threatened by the class struggle. For the author, "non-proletarian wage earners therefore have identities and contradictions with both the bourgeoisie and the proletariat. Their social function, in general, is to help reproduce bourgeois social relations" (LESSA, 2007, p.180).

Thomaz Junior (2009, p. 124) counters this argument by pondering that:

> Two reminders of history are in order. Firstly, if there are plenty of examples to back up this assertion, just as there are plenty to deny it, why doesn't the author consider them, particularly taking into account the current reality in the various corners of the planet? It's not up to us to answer this question, but we do think that the *a priori* definitions that have plagued us so much appear once again - and also through hasty evaluations, the very dynamic content of Marxist dialectics is mutilated by ignoring the negativity and positivity of work. Secondly, limiting the concept of the working class to the identity of the manual, salaried worker, producer of surplus value, or even the broader understanding that recognizes all salaried workers (productive and unproductive) does not take into account the increasingly significant contingents of men and women who do not fit into these premises, and are situated on *the bangs* of supreme exclusion, in other words, those without a job, without work, without a profession, without a home, without land.

In other words, the difficulty in accepting any argument that indicates an explanatory limitation obstructs the debate, so that this intellectual myopia underestimates the necessary rethinking of objective reality and the changes that impose new contradictions and redefine the ontological meanings of social subjects and, consequently, class clashes. What stands out is the unequivocal need to rethink, from within the territorial dynamics of work, the conditions in which class conflict is expressed in society, which necessarily portrays the conflicts and reality of the centrality of work in the 21st century.

In this study we are trying to understand work in its multiple forms of extemalization/pracarization, and the metamorphoses that fall on its universe hit the spatial dynamics of the working migrant being hard, resizing the geographical configuration of territorialization, deterritorialization and reterritorialization, as well as the different forms of expression (material and/or subjective), with profound implications for the heterogenization, complexification and hierarchization of the working class. If, in fact, the question of the working class is a profound dilemma in the theoretical sphere, observing it in practice, in objective reality, is even more complex and confusing.

Returning to the question that drives us, what is the class identity of the migrants who work in the sugarcane fields of Gastâo Vidigal-SP? If, as we said above, they find themselves in conditions of "non-freedom" and controlled by the process of the "alienation of labor", what identities do these subjects postulate and construct during their continuous coming and going?

As Oliveira (2009) states, migrant workers immersed in this confusion or *territorial trance* under the fetishes and alienations mediated by the social metabolism of capital, express forms of identity that are distant from their class belonging, topped by categorical nomenclatures correlated to the technical division of labor.

We can see from the migrants' testimonies that the construction of a working class identity, for migrants working in sugarcane cutting, is fragmented, there is no awareness of class unity, the subjects recognize themselves and bind themselves through the profession they carry out, in these terms, there is a confusion between employment and work.

"Well, I've recognized myself since I was a person, as a *cane cutter*. And, above all, as a cane cutter from "outside", as we are called by the locals". (migrant from Piauí, 26 years old).

In the excerpt above, we can see that the class identity is completely emptied, the subject recognizes himself by the activity he performs, but it doesn't stop there, the socio-territorial issue in the order of regency of the metabolism of capital, also presents itself as an additional obstacle to the unity of the struggle. In other words, as the migrant himself defines himself, "a cane cutter from outside", that is his identity, the links to building a common agenda of demands by workers in this condition are practically shattered.

However, class alienation doesn't just take place in the case of migrants, it is also embodied in the ideals of local workers. As this local resident said, "I see that I'm a cane cutter, but not a migrant". In this case, there is again the confusion between employment and work, and only the recognition of identity with what one does, but the new element in this equation lies in the fact that the "local worker" shows a certain aversion, dread and emphatically repels and ratifies that he is not a migrant worker, as if being a migrant were something shameful.

Therefore, as we can highlight and conclude, the feeling of class belonging is still far from being visualized in the municipality of Gastâo Vidigal-SP, the mediations established by the endorsement of capital do not allow the construction of the unity of the struggle, it does not allow the subjects to recognize in the other the class identity of work, the other is seen only as the other, nothing familiar, it is seen only as a strange being with whom for a certain time we will share the same territory.

A question arises within the migratory movement to Gastâo Vidigal-SP: how do the different subjects hierarchized by the alienating structure of the logic of capital see themselves? What elements shape the relationship between local residents and migrant workers? How does the local population view the migration process?

Roughly speaking, the alienated perception that migration represents a vector of negative effects on the municipal infrastructure stands out, its main claim being that the health service cannot cope with the swelling caused by the migration of 1900 migrants to a municipality of only approximately 4193 inhabitants, see below.

"It's bad because of the health center, because when you go to see a doctor, it's full of people from Piauí and there's no way to see them. And not to mention that the medicines are always missing".

(local resident, 39 years old).

"It's only good for those who are traders, who have a market, to make money off them, it's only good for them. And another thing, if the price goes up to exploit them, it goes up for us here too." (local resident, 23 years old).

Thus, local residents blame the socio-territorial infrastructure problems on the migrant workers, as if they were the ones to blame and responsible for the situation and not the ones who brought them here. In this context, demand also increases for public health services such as hospitals and health centers, since most of the workers depend on these services, since they don't have health insurance or the financial means to pay for care in private institutions. As there is generally little (or no) public investment in this area, the impact also ends up falling on the local population. Thus, care in these public institutions, which is already precarious, with a lack of doctors, long queues, insufficient beds and medicines, tends to get worse.

Based on the testimony of some representatives of the municipal authorities, it is possible to say that the prospect of development created by the construction of sugarcane agro-industrial companies is seen as positive for the municipalities, especially in terms of job creation and tax collection. The interviewees pointed out that it is interesting for the processing plant to be set up in the municipality, as it increases tax collection and allows the government to invest in infrastructure and service improvements. At the same time, it's not interesting for the company to set up in another municipality and demand the services of another, because while the former collects taxes directly linked to the processing plant, the latter absorbs, without compensation, the impacts resulting from the population contingent received, such as demand for housing, health services, security and even education.

As we can see, a considerable portion of the local population sees the migration of migrant workers as an "obstacle" for the municipality, as a negative effect that ends up affecting the "locals". The very language used demonstrates the displeasure of relating to what is different. In general, all migrants' territoriality is homogenized, and in the view of some locals they become simply "Piauienses", referring to any migrant. In other words, the fact that local people identify migrants in a generalized and banal way by the *toponymic* definitions of the origin of these workers, as well as representing prejudice about the physical and ethnic/racial characteristics of these groups that are easily identified by the *locals*, also represents a certain *nostalgia* (or *ufanistic regionalism*) for the supposed superiority of the Southeast region over the Northeast. Therefore, even though all social beings are at the mercy of the terrifying power of capital, a portion of them, imbued in this process of alienation, don't even recognize the human being in the other.

But what about the local rural workers' union? Does it fulfill its historically constructed role? It is a fact that the exploitation of migrant labor in Gastâo Vidigal-SP is not only carried out without the consent of the unions, but also in many cases with their own consent, because the changes in the world of work have led to the emergence of a new model of union, assimilated and co-opted by the companies, which turns their actions into a *neo-corporatism* aimed at preserving the interests of stable, union-bound workers (contributors) to the detriment of migrant workers, *outsiders*, "passengers", temporary workers or casual workers.

The alienation of labor embodied in the unions also expresses itself through a disaggregating action of the unity of action of the world of work, restricting their action to the worker-territory link, i.e. they "represent" a specific category with a corporate identity of a specific territorial base, delimited by the state. As Thomaz Junior (2002, p. 230) shows:

> While capital spreads across the territory, materialized in the form of areas occupied by sugar cane and sugar-alcohol companies, surpassing/subverting the territorial limits imposed by the state (the municipal border), labor, (already in the form of a corporate identity, therefore as a union), has its territorial scope delimited by municipal borders.

In this way, the scope of action for labor as a corporate entity is territorially delimited by the borders of the municipality, as opposed to the territorial spread of capital, which materializes in the sugarcane plantation areas and in the sugar-alcohol companies. Despite being bound by the same legal rules as the workers in terms of representative bodies, capital is unified around a single entity. What's more, as it becomes hegemonic over the production process, it exercises its own management, recognizing itself not within the limits of the municipalities, but in the materialization of production. This corporate union structure, in general, hinders the collective participation of workers, inhibiting the struggle for better working conditions, and "fragments union action, restricting it to the categories represented by the union, not expanding these struggles even when the demand is expanded by the workers, as in the case of wage campaigns,[...]" (CARVALHAL, 2000, p.100).

In this regard, it's worth pointing out that the fragmented management of the various categories of workers and their representative bodies has its origins in the vertical confederative nature of the structured union model in Brazil, based on the union-federation-confederation order. In the sugarcane agro-industry, in particular, this fragmentation is expressed in rural workers (cane cutters), workers linked to the manufacturing process (chemical and food) and workers in the transport sector of the companies (drivers, tractor drivers, etc.), so that work is fractionated within the category, thus reducing unity and the ability to fight.

In this respect, the territorial cleavages of labour are established. This is the high point of the alienation of labour, because while capital subverts barriers and borders to territorialize itself, trade unions are increasingly fragmented and territorially subsumed to its interests. In this way, the territorial ordering of union practice is expressed in the fragmentation of workers between *locals* and *outsiders*. By representing only permanent workers (taxpayers) to the detriment of migrant, seasonal and temporary workers, unions distance themselves from their collective class identity as workers and assume a piecemeal and segmented identity as cane cutters, fixed or permanent and temporary workers, *locals and outsiders*, migrants and non-migrants, etc.

We therefore agree with Thomaz Junior (2009) on the need to think about the intertwining of these struggles in order to move towards overcoming mediatization and atomization, i.e. a counter-hegemonic, anti-capitalist struggle that is organically unified, i.e. capable of going beyond the boundaries of the technical and territorial division of labour, beyond the limits of urban and rural areas. In other words, the challenge is to "subvert" the framework of corporatist fractionalization imposed until now by capital and legitimized by the state. A subversion that could manifest itself through the organic unification of the unions, in other words, the formation of a union entity that prioritizes the rooting of all productive activity and that covers all the workers linked to the productive chain, who today are divided into various categories.

FINAL CONSIDERATIONS

"Everything that is [only]read falls apart in the air"

As an end, a beginning, thinking about the situation of the real heroes of this immense "sea of green" that is territorialized in Gastâo Vidigal-SP, the rural workers, especially migrants, we defend the idea that they are in conditions of "non-freedom", that the act of migrating does not imply a free movement through space, a free will of the worker to migrate, but that migration in the capitalist system is configured as a *forced spatial displacement*.

Because, under the reign of capital's second-order mediations, migrants have their "territory of consciousness" alienated and alienated, becoming a territory for use by the other, in this case agribusiness. It is through this process of conversion and appropriation of territories that agribusiness uses migrant labor as the basis for its extreme gains. Capital in the 21st century isn't satisfied with just extracting *surplus value*, it requires more, and this *plus* is obtained through the process that we detected within migration to cut sugarcane in the municipality of Gastâo Vidigal-SP, through the extreme precariousness of working relationships that are legitimized by the coercive support of the establishment of the "territory of alienation". Because, according to the migrant workers' accounts, they live and feel the punishment of work, they perceive exploitation through the sphere of wealth distribution, which gives concrete form to poverty, but they don't identify its social reasons, related to the capitalist form.

Contrary to what some scholars think, the process of worker migration does not mean a mere demographic rebalancing between territories. We need to go beyond what is visible and understand that behind this forced displacement there are men and women who risk their lives in order not to die of hunger. Therefore, in the face of the ideological onslaught of agribusiness, which presents itself as the salvation of agriculture and the path to development, it is necessary to pay attention to the structural conditions of the capital system that generate the drama of migration.

The first of these is the *concentration* of *land*, wealth and power. Scholars of Brazilian society never tire of highlighting the three pillars on which Brazil's economy has been based since colonial times: large landholdings, export monoculture and slave labor. Throughout history, this concentration has only increased, creating an elite that accumulates a huge share of the national income and has proved to be extremely retrograde and averse to any transformation. Of course, this state of affairs is at the root of many migratory movements. In this sense, land is a *living issue* for migration, a central issue, in which case the fight against all types of forced migration becomes decisive. This brings us to the struggle for land!

In these terms, we agree with Oliveira (1986) that the struggle for land cannot be restricted only and specifically to the struggle for the right of access to land; it must be the struggle against those behind capitalist ownership of land, that is, capital.

Supporting movements in the countryside and efforts for agrarian reform and an effective agricultural policy is undoubtedly a way of preventing thousands of families from leaving compulsorily. Keeping people on the land, with real living and working conditions, is a way of reducing the mass movement of migrants. Without ever forgetting the freedom to come and go, because neither settlement nor migration can be forced.

And finally, the second relates to the precariousness of the world of work, because in Brazil, as in the rest of the world, capitalism today reveals an enormous contradiction: at the same time as it develops the most advanced technology, it resurrects forms of work that have been condemned and consigned to the dustbin of history. This is the case, to cite a few examples, with slave labor, child labor, home-based work, women's work for lower pay, temporary work, piecework *and freelancing.* Terms such as flexibilization and outsourcing are windows into this process which has made working relationships more precarious, reducing stable jobs and multiplying "odd jobs", and in this situation, once again, the worker is forced into a compulsory shuttle.

This is because, as Antunes (2005) points out, under capital's system of social metabolism, the work that structures capital de-structures social being. This is why it becomes more than essential to emphasize the idea of the incorrigibility/irreformability/incontrollability of capital. In other words, since capital is one of the most powerful totalizing control structures, within which everything, including human beings, must fit, prove its productive viability, otherwise it will be extinguished. This makes us think that capital's inability to rise above the *short-term* perspective, the revived destructive competition between capitals, emphasizes that it is a controlling force, and there is no way to control it. Only the possibility of getting rid of it by transforming the whole complex of metabolic relations in society could exterminate it.

In these terms, the clues provided by Mészàros (2003) ring categorically true, in the sense that it is absolutely inconceivable to overcome any of these contradictions, let alone this inextricably combined network, without instituting a radical alternative to the mode of control of capital's social metabolism. In this way, *radical change* is possible, we're not talking about reforms, because as we know, the system of capital metabolism is incorrigible, uncontrollable, there's no way to make capital less evil, because from its roots, from top to bottom, throughout its structure, it is already born sacred and using the expedient of the exploitation of man by man.

Therefore, despite the heterogenization, complexification and fragmentation of the working class, the possibilities of effective human emancipation from the bonds of capital can still find concreteness and social viability from the revolts and rebellions that originate centrally (and not exclusively) in the world of work, A process of emancipation simultaneously from work, in work, and through work, always remembering that this formulation does not exclude or suppress other important forms of rebellion and contestation. This obliges us today, more than ever, to radically subvert the destructive logic of capital that currently presides over humanity, taking it to the deepest levels of inhumanity. And that only humanity can transform!

Returning to the object of our research, it is possible to indicate that as long as the sugarcane sector remains under the rule of capital, there will remain an internal dichotomy in which, on the one hand, the most modern technology is used, but on the other, extremely archaic, barbaric work relations are maintained, in the sense of the regression of the social being in the midst of "modernity".

The impressive growth in labor productivity will only have a humanist destiny if it is socialized, shared by all.

What we are seeing is that sugarcane agribusiness, this clear and current expression of the latent stage of barbarism in which we live, in combination with a conjuncture that exalts neoliberal presuppositions, has been characterized by a strong destructive character, bringing about, among many harmful aspects, an enormous precariousness of work and a growing degradation in the metabolic relationship between man and nature. This combination, driven by a societal logic that is driven by the interests of capital, focuses primarily on the production of goods for the foreign market, at the cost of the destruction of peasant ways of life, forced displacement, the death of workers and the local "environment". Thus, from a *"dream of freedom"*, the reality that is actually being imposed is *migration to capital,* i.e. sugarcane agribusiness, in other words, the *dream turns into a nightmare!*

At this point, we need to make it clear that we are opposed to the enterprise of the capital system and its destructive logic, which more than ever points to the destruction of the human race. Considering this serious structural crisis of the capital system, the inflexible alternative today is *"socialism or barbarism"*, as signaled by Mészàros (2007). Faced with this "historical dilemma", it becomes imperative and an unavoidable historical challenge to build and collectively elaborate another alternative socio-metabolic order that eradicates the destructive and self-expanding logic of production and is structured according to the basic paradigm of substantive equality between human beings and the satisfaction of all their needs.

REFERENCES

ALVES, F. The new dynamics of the sugarcane agro-industrial complex: precarious rights, concentration and social exclusion. In: NOVAES, J. & ALVES, F. **No eito da cana**: Exploraçâo do trabalho **e** luta por direitos na regiâo de Ribeirâo Preto, Sâo Carlos: Rima Editora, 2003. p. 78-81.

. **Why do cane cutters die**?, 2006. Available at: www, adital. com. br/site/tema.asp?lang=EN&cod=23&currpage=4. Accessed on: October 10, 2008.

. Migration of rural workers from Maranhâo and Piaui to cut sugarcane in São Paulo: Is this a random phenomenon or a recurring business strategy of the sugarcane agro-industrial complex? In: NOVAES, R. & ALVES, F. (Orgs.). **Migrants: Trabalho e Trabalhadores no Complexo Agroindustrial Canavieiro (os heroes do agronegócio brasileiro),** Sâo Carlos: EDUFSCar, 2007. p. 21-54.

. Labor and Workers in Sugarcane Cutting: still the controversy over payment for production and deaths from overwork In: SILVA, M. A de M; ALVES, F; PEREIRA, J. C. A. **Agrocombustiveis Soluçâo? a vida por um fio no eito dos canaviais.** Sâo Paulo: Youth Training Center, 2008. p. 22-48.

ANDRADE, B. Coord. **Migraçôes Sazonais no Nordeste** (relatório de pesquisas). Recife, SUDENE, Federal University of Pernambuco. 1981.

ANTUNES, R. **Goodbye to Work?** An Essay on the Metamorphosis and Centrality of the World of Work. 7. ed. Sâo Paulo: Cortez/Unicamp, 2000.

. **The snail and its shell.** Sâo Paulo: Boitempo, 2005.

BACKES, T. R. From the homogenization of the landscape to the expanded reproduction of capital: an analysis of the expansion of sugar cane in the state of Mato Grosso do Sul. In: **Caminhos de Geografia**, Uberlândia, v 9, n° 26, 2008.

BASBAUM, L. **Alienation and humanism**. Sâo Paulo: Fulgor, 1967.

BRAZIL. Ministry of Industry and Trade. **National Alcohol Program**, Brasilia, 1986.

CARDOSO, M. A. **Migraçâo para o Capital nos Canaviais do Oeste Paulista: The Geography of Work and the Imperatives of Barbarism**, In: X Jornada do Trabalho: The importance of theory for social transformation and the indispensability of research for the materialization of the emancipatory praxis of the working class in the 21st century, Pres. Prudente, Anais Eletrônicos, 2009 Available at: <http://www4.fct.unesp.br/ceget/anaisxjornadatrabalhos.htm> Accessed on: 26 Nov. 2009.

CARVALHAL, M. D. **A comunicaçao sindical em Presidente Prudente/SP**: elementos para uma leitura geogràfica. 2000, 168 f. Dissertation (Master's in Geography) - Faculty of Science and Technology, Paulista State University, Presidente Prudente, 2000.

DANIEL, et al. **Characterization of the expansion of sugarcane cultivation and the types of contracts in the Jales and General Salgado regions of the State of São Paulo.** In: 47th Congress on Rural Development and Agrifood Systems: Agribusiness in the Context of Nation Integration. Porto Alegre, 2009.

FERNANDES, B. M. Entering the territories of the Territory. In: **Campesinato e Territórios em Disputa**. Sâo Paulo: Expressâo Popular, 2008.

On the Typology of Territories. In: Saquet, Marco Aurélio; Sposito, Eliseu Saverio. (Org.). Territories and territorialities: theories, processes and conflicts. Sâo Paulo: Expressâo Popular, 2009.

. The agrarian question: conflict and territorial development. In: BUAINAIN, A. M. Editor. **The Struggle for Land, Agrarian Reform and Conflict Management in Brazil**. Campinas: Unicamp, 2004.

GAUDEMAR, J-P. de. **Mobility of labor and accumulation of capital**. Translated by Maria de Rosàrio Quintela. Lisbon: Editorial Estampa, 1977.

GOMES, F. G. **Mobilidade do trabalho e controle social**: **trabalho e organizaçoes na era neoliberal**. Revista Sociologia Politica, Curitiba, vol.17, n.32, p. 33-49, February. 2009.

GONÇALVES, A. J. Migraçôes Internas: evoluções e desafios. **Estudos Avançados** (15), 2001.

GUANAIS, J. B. Impasses of the work process in the sugarcane agro-industry: manual cutting of sugarcane and payment for production In: SILVA, M. A de M; ALVES, F; PEREIRA, J. C. A. **Agrocombustiveis** Soluçâo**? a vida por um fio no eito dos canaviais**, Sâo Paulo: Centro de Capacitaçâo da Juventude, 2008. p. 49-57.

IAMAMOTO, M. V. **Trabalho e Individuo social**: um estudo sobre a condiçâo operadora na agroindùstria canavieira paulista. Sâo Paulo, Cortez, 2001.

. O trabalho **no** canavial: sofrimento **e** rebeldia na luta pela defesa **da** vida In: NOVAES, J. & ALVES, F **No eito da cana**: Exploraçâo do trabalho **e** luta por direitos na regiâo de Ribeirâo Preto, Sâo Carlos: Rima Editora, 2003. p. 69-71.

IANNI, O. **A luta pela terra**. 3rd ed. Petrópolis: Vozes, 1981.

LESSA, S. **Trabalho e proletariado no capitalismo contemporâneo**. Sâo Paulo: Cortez, 2007.

MONTENEGRO GOMEZ, J. R. **Desenvolvimento em (des)construçâo:** Narrativas Escalares Sobre Desenvolvimento Territorial Rural. Thesis (Doctorate in Geography). Faculdade de Ciências e Tecnologia de Presidente Prudente, Universidade Estadual Paulista, Presidente Prudente, 2004.

MARTINS, J. S. **Peasants and politics in Brazil.** Social struggles in the countryside and their place in the political process. Vozes, 2 ed. Petrópolis, 1983.

. **The power of delay**. Essays on the Sociology of Slow History. Sâo Paulo: Hucitec, 2001.

MENEGAT, M. The face and the mask: the barbarism of bourgeois civilization. Presidente Prudente, **Pegada**, V.8, N.2, 2007. p. 27-46.

. **The eye of barbarism**. Sâo Paulo: Expressâo Popular, 2006.

MÉSZAROS, I. **O desafio e o fardo do tempo histórico**: o socialismo no século XXI, transl. Ana Cotrim, Vera Cotrim, Sâo Paulo: Boitempo, 2007.

. **The 21st century: socialism or barbarism?** Sâo Paulo: Boitempo, 2003.

MOREIRA, R. **Pensar e ser em geografia: ensaios de história, epistemologia e ontologia do espaço geogràfico.** 1ª ed. 1ª reprint. Sâo Paulo: Contexto, 2008.

OLIVEIRA, A. U de. **Capitalist Mode of Production and Agriculture**. Sâo Paulo: Atica, 1986.

OLIVEIRA, A. M. S. de. **Territorial and productive reorganization of sugarcane agribusiness in Brazil and the consequences for work**. PhD thesis- UNESP, Presidente Prudente, 2009.

. The new frontiers of sugarcane agribusiness in Brazil: weaving insights into working conditions and food security. Presidente Prudente, **Pegada**, V.8, N.1, 2007. p. 47-75.

. ______The Territorialization of Sugarcane Agroindustrial Capital and the *New* Geography of Migrant Labor in Brazil. In: THOMAZ JÙNIOR, A; GONÇALVES, M. A; OLIVEIRA, A. M. S. de. (Orgs.). **Geography and Work in the 21st Century,** Volume III. Presidente Prudente, 2007. p. 54-83.

PEREIRA, J. C. A. The place is good for living, but the conditions don't make it so. In: SILVA, M. A de M; ALVES, F; PEREIRA, J. C. A. **Agrocombustiveis** Soluçâo**? a vida por um fio no eito dos canaviais**, Sâo Paulo: Centro de Capacitaçâo da Juventude, 2008. p. 58-71.

PERPETUA, G. M. Commuting and capital accumulation. **Revista Pegada Eletrônica**, Presidente Prudente, vol. 11, n. 2, December 31, 2010. Available at:

<http://www.fct.unesp.br/ceget/pegada112/07MARINI1102.pdf>. Accessed on: October 20, 2011.

PRADO JUNIOR, C. **O que é Liberdade**: capitalismo x socialismo. Sâo Paulo: Ed. Brasiliense, 12ª Ed., 1986.

SEN, A. **Development as Freedom**. Trad. Laura Teixeira Motta. Sâo Paulo: Companhia das Letras, 2000.

SILVA, M. A. de M. Rural workers: the denial of rights. **Raizes**. Vol.27 n.1,2008.

. _______Agronegócio: a reinvençâo da colònia In: SILVA, M. A de M; ALVES, F; PEREIRA, J. C. A. **Agrocombustiveis Soluçâ?** a vida por um fio no eito dos canaviais, Sâo Paulo: Centro de Capacitaçâo da Juventude, 2008. p. 4-18.

. Work and Workers in the "Sea of Sugar Cane and Rio de Alcohol" Region In: NOVAES, R. & ALVES, F (Orgs.). **Migrants**: Trabalho e Trabalhadores no Complexo Agroindustrial Canavieiro (os heroes do agronegócio brasileiro), Sâo Carlos: EDUFSCar, 2007. p. 55-86.

. Death prowls the sugarcane fields of São Paulo. Sâo Paulo. **Reforma Agrària**, vol.33- n. 2, 2006, p. 111- 148.

. **Wanderers at the end of the century**. Sâo Paulo: Editora Unesp, 2002.

SILVA, M. A. M.; MENEZES, M. A. **Temporary Migrants: The End of Narrators?** NEHU/USP, n° 1, 2000.

SINGER, P. Migraçôes Internas: Consideraçôes teóricas sobre o seu estudo. In: **Economia Politica da Urbanizaçao**. 6. ed. Sâo Paulo: Brasiliense, 1979, p. 29-60.

SOUZA, S. M. R. de. **The expansion of sugarcane activity in the state of São Paulo**: the country's locomotive and the portrait of backwardness in labor relations In: IX Jornada do Trabalho Dinàmica Territorial do Trabalho no

Século XXI: em busca dos sujeitos que podem emancipar a sociedade além do capital, Catalâo-GO, Anais Eletrônicos, 2008.

THOMAZ JÛNIOR, A. **Nova territorialidade de luta do movimento sindical dos trabalhadores do CAI sucro-alcooleiro em Sao Paulo (reflexôes iniciais),** 1992 (Text).

. **Behind the sugarcane fields, the "knots" of sugarcane**: the relationship between capital and labor and the workers' union movement in São Paulo's sugarcane agroindustry. Sâo Paulo: Annablume/FAPESP, 2002.

. **Alcoholized Agribusiness and Expanding Crops in Pontal do Parapanema! Legitimization of Vacant Lands and Neutralization of Social Movements.** Presidente Prudente, 2007 (mimeo).

. There is nothing new under the sun in a world of heroes! (The civilization of barbarism in the sugarcane agroindustry). Presidente Prudente, **Pegada**, V. 8, N.2, 2007. p. 5-25.

. **The Geographical Dynamics of Work in the 21st Century (explanatory limits, self-criticism and theoretical challenges)** (Free Teaching Thesis - UNESP) Presidente Prudente, 2009.

. The Destructive Development of Productive Forces, the Unsustainability of Capital and the Challenges for Food Production. In: THOMAZ JUNIOR, A.; FRANÇA JUNIOR, L. B. **Geografia e Trabalho no Século XXI,** volume V. Presidente Prudente: Editorial Centelha, 2010. pp. 176216.

VALE, A. L. F **Migration and Territorialization**: the territorial dimensions of Northeasterners in Boa Vista/RR. PhD thesis- UNESP, Presidente Prudente, 2007.

WELCH, C. A; FERNANDES, B. M. **Agricultura e mercado: campesinato e agonegócio da laranja nos EUA e Brasil**. Presidente Prudente; Digit: 2008.

MIX
Papier aus verantwortungsvollen Quellen
Paper from responsible sources
FSC® C105338

Printed by Books on Demand GmbH, Norderstedt / Germany